SpringerBriefs in Research and Innovation Governance

More information about this series at http://www.springer.com/series/13811

Doris Schroeder · Julie Cook
François Hirsch · Solveig Fenet
Vasantha Muthuswamy
Editors

Ethics Dumping

Case Studies from North-South Research
Collaborations

Editors
Doris Schroeder
Centre for Professional Ethics, School
   of Health Sciences
University of Central Lancashire
Preston, Lancashire
UK

Julie Cook
Centre for Professional Ethics
University of Central Lancashire
Preston, Lancashire
UK

François Hirsch
Institut National de la Santé et de la
   Recherche Médicale
Paris
France

Solveig Fenet
Institut National de la Santé et de la
   Recherche Médicale
Paris
France

Vasantha Muthuswamy
Manchester Regent
Coimbatore, Tamil Nadu
India

ISSN 2452-0519   ISSN 2452-0527   (electronic)
SpringerBriefs in Research and Innovation Governance
ISBN 978-3-319-64730-2   ISBN 978-3-319-64731-9   (eBook)
https://doi.org/10.1007/978-3-319-64731-9

Library of Congress Control Number: 2017953819

Printed on acid-free paper

This Springer imprint is published by Springer Nature
The registered company is Springer International Publishing AG
The registered company address is: Gewerbestrasse 11, 6330 Cham, Switzerland

To Andries Steenkamp (1960–2016)

Foreword

Today, humanity faces grand challenges of a global nature, such as epidemics, disease outbreaks, food security challenges, climate change and increasing energy demands. Such challenges can only be resolved through cooperation between countries, as no country acting alone could possibly tackle them effectively.

A main driver towards the resolution of these challenges is research and innovation. International cooperation in research and innovation is particularly important, as it promotes the sharing of knowledge, skills and resources; it raises awareness and may contribute to the fair sharing of the benefits of research and innovation among partners. Undertaking international cooperation in research and innovation *responsibly* requires equitable and respectful relationships between countries and among research and innovation partners. Fair research relationships are particularly important for Europe, given its leading roles in responsible research and innovation and human rights compliance, in order to foster the building and maintaining of equitable research partnerships with low- and middle-income countries (LMICs).

When it comes to international cooperation with LMICs, the most fundamental element of responsible research and innovation is adherence to high ethical standards, independently of where the research takes place. This is particularly the case when we take into account the power imbalances and disparities in know-how between high-income countries and LMICs, which may result in the former taking advantage of the vulnerabilities of the latter. Participants and resources from LMICs must not be exploited in international partnerships, even if local ethics compliance structures are weak compared with compliance structures in Europe. In other words, research and innovation partners should refrain from taking either active or passive advantage of loopholes and weaknesses in the governance systems of another country in order to perform research that would be legally or ethically unacceptable in their own country.

Ensuring equitable and respectful partnerships and avoiding exploitation of LMICs require mutual understanding. In order to reach such mutual understanding, it is of particular importance to examine case studies, as they may elucidate ethical issues relevant to real-life recent research and innovation activities involving

LMICs, which may contribute to avoiding the duplication of past mistakes and injustices.

This book aims to raise awareness of the topic of unethical research and therefore presents case studies of exploitative research conducted in LMICs. Funded by the European Commission, it brings together experts on this topic from around the world. Adhering closely to an important feature of responsible research and innovation, namely societal engagement, the book has directly involved highly vulnerable populations in its outputs (LMIC sex workers and indigenous peoples).

In order to maximize outreach, this book is made available as gold open access. This means that stakeholders from LMICs who have access to the Internet will not be excluded from the learning outcomes of this work. Learning about the exploitation of LMICs from open-access case studies will hopefully increase the chances that those involved in research will be open to the world in an inclusive and equitable manner. Presenting such case studies aims to contribute to the development of fair and equitable research and innovation collaborations between countries —collaborations that are mutually beneficial for all participants, as well as for scientific progress at large.

Louiza Kalokairinou is a Policy Officer at the Ethics and Research Integrity Sector of the Directorate General (DG) for Research and Innovation (European Commission), and a Ph.D. candidate at the Centre for Biomedical Ethics and Law (University of Leuven).

Isidoros Karatzas is Head of the Ethics and Research Integrity Sector, European Commission (EC), DG Research & Innovation. A biochemist by training, he established advanced training courses on research ethics and research integrity for EC staff, the research ethics expert community, early career researchers and National and European professional associations. He was the first to set up a European system of ethics checks and contributed to the publication of the new European Code of Conduct for Research Integrity and the preparation of the Horizon 2020 research integrity strategy.

Acknowledgements

This book is an output of the TRUST project funded by the European Commission.[1] We first want to thank our funder, and in particular Mr. Dorian Karatzas, head of the Ethics and Research Integrity Sector at the European Commission's Directorate-General for Research and Innovation. It was Dorian's long-standing commitment to ensuring that learning for Horizon 2020 stakeholders takes place through case studies that led to this book.

We also thank our project officer, Roberta Monachello from the Research Executive Agency of the European Commission, for her engaged interest in the TRUST project and Louiza Kalokairinou from the Ethics and Research Integrity Sector for promoting our work. The project was chosen as one of Horizon 2020's success stories (SiS.net nd), which energized us greatly, and we hope that those interested in this book will also find the project's future outputs useful.[2]

The authors of these case studies come from Austria, Cameroon, Canada, China, France, Germany, India, Kenya, Liberia, the Philippines, Russia, South Africa and the UK. We would like to thank them all for bringing these cases to our attention.

The inclusion of a case from India is thanks to Dr. Urmila Thatte and Dr. Vasantha Muthuswamy, who organized an inspirational workshop in Mumbai in 2016. Here, we also thank Dr. Nandini Kumar for her specialist advice.

This is the fifth book that one of the editors, Doris Schroeder, has published with senior Springer editor Fritz Schmuhl. As before, his problem-solving attention from the first email to the last step in production has been invaluable. It is also the fifth collaboration with our copy editor, Paul Wise, who is simply the best.

Various institutional services at the University of Central Lancashire in the UK, where TRUST is coordinated, have supported this book indirectly. In particular, we would like to thank Aisha Malik (senior finance officer), Michelle Cartmell and her team (travel office), Kate Hutchinson (funding development and support) and Alexander Rawcliffe (research support team). Isabelle Pires (accounting at the

---

[1]Project number 664771.

[2]trust-project.eu/.

French National Institute of Health and Medical Research) also provided essential support. Our thanks go to you all.

Lastly, but most importantly, we thank Andries Steenkamp, the San traditional leader from the Kalahari in South Africa, for his leadership over many years in working to make North-South research relationships more equitable. We dedicate this book to his memory.

Reference

SiS.net (nd) Success stories. Network of National Contact Points for Science with and for Society in Horizon 2020. http://www.sisnetwork.eu/about/success-stories/

Contents

1   **Ethics Dumping: Introduction**   1
    Doris Schroeder, Julie Cook, François Hirsch, Solveig Fenet
    and Vasantha Muthuswamy

2   **Social Science Research in a Humanitarian Emergency**
    Context.   9
    Gwenaëlle Luc and Chiara Altare

3   **International Genomics Research Involving the San People**   15
    Roger Chennells and Andries Steenkamp

4   **Sex Workers Involved in HIV/AIDS Research**   23
    Anthony Tukai

5   **Cervical Cancer Screening in India**   33
    Sandhya Srinivasan, Veena Johari and Amar Jesani

6   **Ebola Vaccine Trials**.   49
    Godfrey B. Tangwa, Katharine Browne and Doris Schroeder

7   **Hepatitis B Study with Gender Inequities**   61
    Olga Kubar

8   **Healthy Volunteers in Clinical Studies**.   67
    Klaus Michael Leisinger, Karin Monika Schmitt
    and François Bompart

9   **An International Collaborative Genetic Research Project**
    Conducted in China   71
    Yandong Zhao and Wenxia Zhang

10 **The Use of Non-human Primates in Research**.   81
    Kate Chatfield and David Morton

11 Human Food Trial of a Transgenic Fruit 91
Jaci van Niekerk and Rachel Wynberg

12 ICT and Mobile Data for Health Research 99
David Coles, Jane Wathuta and Pamela Andanda

13 Safety and Security Risks of CRISPR/Cas9 107
Johannes Rath

**14 Seeking Retrospective Approval for a Study in Resource-
Constrained Liberia** 115
Jemee K. Tegli

**15 Legal and Ethical Issues of Justice: Global and Local
Perspectives on Compensation for Serious Adverse Events in
Clinical Trials** 121
Yali Cong

Other Resources 129

Index 131

About the Editors

Doris Schroeder is director of the Centre for Professional Ethics at the University of Central Lancashire and the School of Law, UCLan Cyprus, and adjunct professor at the Centre for Applied Philosophy and Public Ethics, Charles Sturt University, Canberra. She is the coordinator of the TRUST project and has previously guided large international consortia on the topics of benefit sharing and responsible research and innovation.

Julie Cook is a research associate in the Faculty of Health and Wellbeing at the University of Central Lancashire, where she works closely with the Centre for Professional Ethics and is a member of the Research Ethics Committee.

François Hirsch is head of the Office for Ethics at the French National Institute of Health and Medical Research (Inserm) and assistant director for ethics and regulation at the Institute for Health Technologies. François is currently a member of Comité de Protection des Personnes Ile de France VII.

Solveig Fenet is a researcher at the French National Institute of Health and Medical Research (Inserm). Solveig was previously an economic analyst at the French Development Agency.

Vasantha Muthuswamy recently retired as senior deputy director general of the Indian Council of Medical Research, New Delhi. She was chief of the ICMR's Division of Basic Medical Sciences, Traditional Medicine and Bioethics and chief of the Division of Reproductive Health and Nutrition. A WHO Fellow at the Kennedy Institute for Ethics, Georgetown University, Washington, DC, she is internationally recognized for publishing the ICMR's Ethical Guidelines for Biomedical Research on Human Subjects in 2000 and the revised Ethical Guidelines for Biomedical Research on Human Participants in 2006. She is currently president of the Forum for Ethics Review Committees in India.

Chapter 1
Ethics Dumping: Introduction

**Doris Schroeder, Julie Cook, François Hirsch, Solveig Fenet
and Vasantha Muthuswamy**

Abstract Achieving equity in international research is a pressing concern. Exploitation in any scenario, whether of human research participants, institutions, local communities, animals or the environment, raises the overarching question of how to avoid such exploitation. Agreed principles can be universally applied to research in any discipline or geographical area, whatever methodologies are employed. This chapter introduces a collection of case studies, presenting a range of up-to-date examples of exploitation in North-South research collaborations, in order to raise awareness of ethics dumping.

Keywords Research ethics · Responsible research and innovation
Ethics dumping · North South collaborations · Exploitation

Introduction

Achieving equity in international research is a pressing concern. Exploitative North-South research collaborations often follow patterns established in colonial times. Whether the objects of exploitation are human research participants, institutions, local communities, animals or the environment, this raises questions about how such exploitation can be avoided.

"Dumping" is a term used in economics to describe predatory pricing policies in international trade (Investopedia nd). Dumping usually involves substantial export

D. Schroeder (✉) · J. Cook · F. Hirsch · S. Fenet · V. Muthuswamy
Centre for Professional Ethics, University of Central Lancashire,
Brook 424, PR1 2HE Preston, UK
e-mail: dschroeder@uclan.ac.uk

© The Author(s) 2018
D. Schroeder et al. (eds.), *Ethics Dumping*, SpringerBriefs in Research
and Innovation Governance, https://doi.org/10.1007/978-3-319-64731-9_1

1

volumes of a product and often has the effect of endangering the financial viability of manufacturers of the product in the importing nation.

"Ethics dumping"[1] occurs mainly in two areas. First, when research participants and/or resources in low- and middle-income countries (LMICs) are exploited *intentionally*, for instance because research can be undertaken in an LMIC that would be prohibited in a high-income country. Second, exploitation can occur due to insufficient ethics awareness on the part of the researcher, or low research governance capacity in the host nation.

This book provides 14 case studies of ethics dumping and one case of good practice. Its purpose is to address the second cause of ethics dumping by reducing researchers' lack of awareness.

Background to Ethics Dumping

Jeffrey Sachs, one of the world's leading experts on economic development, noted:

> Technology has been the main force behind the long-term increases in income in the rich world, not exploitation of the poor. That news is very good indeed because it suggests that all of the world ... has a reasonable hope of reaping the benefits of technological advance (Sachs 2005: 31).

It is essential that the progress of science and technology is not accompanied by reasonable claims of exploitation of the poor and vulnerable. This is not easy to achieve, as both moderate poverty[2] and extreme[3] poverty increase the likelihood that communities and individuals will be exploited.

Unevenness in ethical and legal standards has led to the exploitation of human research participants and resources in LMICs that could have been avoided. The international debate on bioethics has noted the existence of "double standards" (Macklin 2004).

Vulnerable populations and research participants worldwide have been protected for decades by research ethics committees (ECs), but their success depends on three conditions. First, a relevant EC must exist with the capability, resources and independence to deal with ethics applications. Second, such committees must be able to recognize culturally sensitive ethical issues in complex settings. Third, a

---

[1]The term was introduced by the Science with and for Society Unit of the European Commission: "Due to the progressive globalisation of research activities, the risk is higher that research with sensitive ethical issues is conducted by European organisations outside the EU in a way that would not be accepted in Europe from an ethical point of view. This exportation of these non-compliant research practices is called ethics dumping" (European Commission nd).

[2]Households can only just meet basic needs for survival, with little left for the education of their children.

[3]Households cannot meet basic needs for survival (e.g. chronic hunger, no access to health care).

compliance mechanism must be in place. As these conditions cannot be guaranteed in LMICs, there is always the risk of an implementation gap.

The first condition (a capable EC) cannot be taken for granted, as in this list of constraints on African ECs:

- Insufficient resources
- Lack of or insufficient expertise on ethical review
- Pressure from researchers
- Lack of active or consistent participation of EC members
- Lack of recognition of the importance of EC functions
- No or poor support from the EC's institution
- Lack of independence
- Pressure from sponsors
- Unequal treatment of applicants in review (Nyika et al. 2009: 193)

The importance of cultural sensitivity is demonstrated in Chap. 4, which describes a study that was granted ethics approval in both a high-income *and* a middle-income country, but failed to consider culturally relevant ethical concerns. The third condition (a compliance mechanism) exceeds the remit of this book, but will be considered further in the TRUST project.[4]

The Cases[5]

Cases of exploitation in research have been used to illustrate unacceptable practices since the mid twentieth century. However, infamous medical experiments, as cited in many textbooks—for example, diabolical Nazi experimentation and the Tuskegee study (Emanuel et al. 2011)—are not always a suitable sole learning source for twenty-first-century researchers.

The case studies in this book will help researchers understand better how exploitation can occur in the context of contemporary North-South collaborations. These are genuine cases, assembled from four sources. TRUST experts contributed case studies. Two non-governmental organizations (NGOs) each contributed a case study. Indian bioethicists were invited to a workshop in Mumbai in 2016 to share their ideas, and a case study competition launched through TRUST sourced additional material from LMICs.

---

[4]http://trust-project.eu/.

[5]Responsibility for the accuracy of each case study, the integrity of the information cited and the legitimacy of its acquisition rests with the respective authors. This disclaimer is especially relevant to those cases where the editors could not verify publicly available sources.

The selected case studies have been grouped into six themes:

- Vulnerable populations
- Clinical trials
- Benefit sharing
- Animal research
- New and emerging technologies
- Ethical governance and processes

Vulnerable Populations

"Social Science Research in a Humanitarian Emergency Context", by Gwenaëlle Luc and Chiara Altare, describes conflicts for an international NGO in an African village. The community felt betrayed when unexpected findings about health-seeking behaviours that revealed illegal female genital mutilation (FGM) were shared publicly and contributed to cultural stigmatization. The NGO performed a dual role as assistance provider and researcher, which endangered the neutrality of the data collection and, in the end, the acceptability of its assistance.

Roger Chennells and Andries Steenkamp criticize an international research project, which aimed to examine the genetic structure of "indigenous hunter-gatherer peoples" from Namibia and compare the results with "Bantu from southern Africa". A supplementary document published with the study contained conclusions and details that the San regarded as pejorative and discriminatory; "International Genomics Research Involving the San People" details the perceived exploitation and the San response.

In "Sex Workers Involved in HIV/AIDS Research", Anthony Tukai tells the personal story of supporting a vulnerable and stigmatized population in a Nairobi slum. In a demonstration of good practice, the case outlines empowerment mechanisms that reduced the potential for exploitation.

In "Cervical Cancer Screening Trials on Poor and Illiterate Women in India", Sandhya Srinivasan, Veena Johari and Amar Jesani describe three internationally funded clinical trials that took place between 1998 and 2015 to determine whether primary healthcare workers could conduct cervical cancer screening using cheap visual inspection. These non-drug trials did not require regulatory permission, and the existing standard of care was misconstrued. According to the authors, known and effective methods of cervical cancer screening (by Pap smear) were withheld from 141,000 women even though they have represented the standard of care in India since the 1970s. Two hundred and fifty-four women in the no-screening arm died from cervical cancer.

Clinical Trials

Godfrey Tangwa questions clinical trials in "A Match to Local Health Needs? Ebola Vaccine Trials". The Ebola epidemic of 2013 in West Africa which affected three countries had been brought under reasonable control by 2015. This case study is about a phase I/II clinical trial (testing for safety and immunogenicity) of a candidate Ebola virus vaccine in 2015 in a sub-Saharan country which had not registered any cases of Ebola. The study was sponsored and funded by one of the biggest northern multinational pharmaceutical companies and had government support. But public concerns about the risk of a public health disaster meant the trial was suspended. A commentary by Katharine Browne and Doris Schroeder discusses the importance of trust, highlighting differences from a 2014 phase I Ebola vaccine trial in Canada.

In "Hepatitis B Study with Gender Inequities", Olga Kubar explores why a proposed internationally sponsored study in Russia was not approved by the local EC. Indications of exploitation consisted of inadequacies in the study's design compared with its announced purpose and the indirect inclusion of women in the trial without their informed consent. On the basis of non-compliance with national and international regulatory and ethical requirements, this trial was not approved, providing an example of successful research ethics governance.

In resource-limited settings, healthy volunteers are most often poor people with low literacy levels who might not understand the risks they are taking, and are in no position to refuse even small financial incentives. Participation in clinical trials is a critical source of income, and some volunteers covertly enrol in several studies simultaneously. This exposes them to medical risks (e.g. drug-drug interactions) and also potentially biases the study data; "Healthy Volunteers in Clinical Studies", by Klaus Leisinger, Karin Schmitt and Francois Bompart, provides a recommendation to protect healthy volunteers from such exploitation.

Benefit Sharing

In "An International Collaborative Genetic Research Project Conducted in China", Yandong Zhao and Wenxia Zhang describe how US university researchers collected blood samples from villagers with the cooperation of local research institutes and the government. The US team was later accused of violating research ethics principles by not adequately informing participants and not sharing benefits fairly. Subsequent investigations by American and Chinese media and authorities showed that the US research institute, its personnel and a pharmaceutical company were benefiting substantially from the project, while the Chinese research participants and the government were not.

Animal Research

In "The Use of Non-human Primates in Research", Kate Chatfield and David Morton show that since regulations on the use of non-human primates are tight in the European Union the number used has declined. However, the increase in numbers used elsewhere indicates that researchers from high-income countries are taking advantage of variations in standards, legislation and humane practices to conduct experiments through collaborative efforts in countries where regulation is less strict.

New and Emerging Technologies

Jaci van Niekerk and Rachel Wynberg present concerns about research to develop a genetically modified "vitamin-enriched" banana for cultivation in Uganda through a proposed trial with North American university students. "Human Food Trial of a Transgenic Fruit" explains how northern researchers and philanthropic organizations determine research priorities without necessarily involving affected LMICs. The case highlights differences between the concepts of food security and food sovereignty, illuminating different approaches to addressing poverty-induced hunger and malnutrition.

"mHealth" is the application of mobile phones or other remote monitoring devices to health care. Mobile phones that can run software applications are increasingly used to improve diagnosis, personalize care and expand access to information and services. But mobile phones also collect a wide range of personal information from users. In "ICT and Mobile Data for Health Research", David Coles, Jane Wathuta and Pamela Andanda focus on the potential ethical issues as researchers and clinicians attempt to minimize unintended harms in new digital territory.

Johannes Rath describes "Safety and Security Risks of CRISPR/Cas9" and other novel genome editing technologies. The case focuses on the unresolved ethical issues related to safety and security in the proliferation of a new and very powerful technology at a time when tailored ethical and legal frameworks at the international, national and local levels are missing.

Ethical Governance and Processes

In "Seeking Retrospective Approval for a Study in Resource-Constrained Liberia", Jemee Tegli describes an attempt to seek ethics approval for an anthropological study after it had been conducted. "Emergency research" was used as a cover to avoid the review process, although emergency research regulations stipulated full disclosure of proposed research prior to implementation.

In "Legal and Ethical Issues of Justice: Global and Local Perspectives on Compensation for Serious Adverse Events in Clinical Trials", Yali Cong analyses a situation in which a major international pharmaceutical company sponsored clinical research in an LMIC and applied a double standard in dealing with serious adverse events (SAEs). A 78-year-old Chinese woman joined a clinical trial, and the sponsor paid the cost of medical care arising from an SAE, but refused the family's request for compensation. The family sued the company and the hospital in litigation that continued for nine years.

The editors of this collection hope that it contributes to raising awareness about the dangers of ethics dumping and unethical conduct in North-South research collaborations and promotes ever higher ethical standards in research conducted anywhere in the world.

References

Emanuel EJ, Grady C, Crouch RA, Lie RK, Miller FG, Wendler D (eds) (2011) The Oxford textbook of clinical research ethics. Oxford University Press, Oxford

European Commission (nd) Ethics. Horizon 2020: the EU framework programme for research and innovation. https://ec.europa.eu/programmes/horizon2020/en/h2020-section/ethics

Investopedia (nd) Dumping. Investopedia. http://www.investopedia.com/terms/d/dumping.asp

Macklin R (2004) Double standards in medical research in developing countries. Cambridge University Press, Cambridge

Nyika A, Kilama W, Chilengi R, Tangwa G, Tindana P, Ndebele P, Ikingura J (2009) Composition, training needs and independence of ethics review committees across Africa: are the gate-keepers rising to the emerging challenges? Journal of Medical Ethics 35(3):189−193

Sachs J (2005) The end of poverty. Penguin, London, Penguin, p. 31, our emphasis

Author Biographies

Doris Schroeder is director of the Centre for Professional Ethics at the University of Central Lancashire, and the School of Law, UCLan Cyprus, and adjunct professor at the Centre for Applied Philosophy and Public Ethics, Charles Sturt University, Canberra. She is coordinator of the TRUST project and has previously guided large international consortia on the topics of benefit sharing and responsible research and innovation.

Julie Cook is a research associate in the Faculty of Health and Wellbeing at the University of Central Lancashire, where she works closely with the Centre for Professional Ethics and is a member of the Research Ethics Committee.

François Hirsch is head of the Office for Ethics at the French National Institute of Health and Medical Research (Inserm) and assistant director for ethics and regulation at the Institute for Health Technologies. François is currently a member of Comité de Protection des Personnes Ile-de-France VII.

Solveig Fenet is a researcher at the French National Institute of Health and Medical Research (Inserm). Solveig was previously an economic analyst at the French Development Agency.

Vasantha Muthuswamy recently retired as senior deputy director-general of the Indian Council of Medical Research, New Delhi. She was chief of the ICMR's Division of Basic Medical Sciences, Traditional Medicine and Bioethics, and chief of the Division of Reproductive Health and Nutrition. A WHO Fellow at the Kennedy Institute for Ethics, Georgetown University, Washington, DC, she is internationally recognized for publishing the ICMR's Ethical Guidelines for Biomedical Research on Human Subjects in 2000 and the revised Ethical Guidelines for Biomedical Research on Human Participants in 2006. She is currently president of the Forum for Ethics Review Committees in India.

Chapter 2
Social Science Research in a Humanitarian Emergency Context

Gwenaëlle Luc and Chiara Altare

Abstract This case study about research in an emergency setting depicts how unexpected findings created conflicts of conscience for non-governmental organization (NGO) workers and exposed research participants and their community to retribution and compromised the local social structure. The community felt betrayed when unexpected findings from research about health seeking behaviours revealing illegal female genital mutilation were shared publicly and contributed to stigmatizing their culture. In addition, the NGO involved performed a dual role – that of assistance provider as well as researcher – which endangered the neutrality of the data collection and, in the end, the acceptability of the NGO as assistance provider.

Keywords Ethics · Female genital mutilation · Unexpected findings in research
Cultural relativity

Area of Risk of Exploitation

This case study covers two potential areas of ethics risks or potential for exploitation.

First, a potential for ethics risks can exist when the ethical standards developed in one context (Western medical research) are applied in another context without due attention to local social norms or communication with local communities. A case can be particularly serious if a local practice violates the laws of the country the research takes place in, as in this case.

Second, a conflict of interest can arise when an assistance provider also conducts research. For instance, this could create expectations among participants, and influence their consent to be enrolled in the study.

G. Luc (✉) · C. Altare
Action Contre la Faim, 1 boulevard de Clichy, 75009 Paris, France
e-mail: gluc@actioncontrelafaim.org

© The Author(s) 2018
D. Schroeder et al. (eds.), *Ethics Dumping*, SpringerBriefs in Research
and Innovation Governance, https://doi.org/10.1007/978-3-319-64731-9_2

9

Background

A major ethical dilemma when conducting research in a volatile emergency setting including culturally heterogeneous groups is the need to balance the risks and benefits for the research participants. An example of such a setting is a refugee camp. Acquiring a clear understanding of context-related risks is challenging: unanticipated risks, if not properly understood or taken into account, could lead to the exploitation of participants or communities.

Research in emergency settings is associated with a range of ethical challenges, as both implementers and participants might be situated in a position of vulnerability and insecurity. In addition, in an emergency setting there may be a need for a rapid response, and it might be difficult for local communities (or the aid providers) to distinguish relief from research, among other things.

In research, the "do no harm" imperative requires that research participants not be put at any additional risk (WMA 2013). This is particularly important in cases where vulnerable participants in emergency settings may not get any direct benefits from the research themselves, but may contribute to producing evidence that will improve interventions with similar populations or in similar settings in the future.

Here we describe a case where research activities did put participants at risk, while simultaneously providing no direct (personal) benefits to them, which led to community complaints. The community felt betrayed because the research did not respond to their needs and priorities, and contributed to stigmatizing their culture.

Specific Case and Analysis

A socio-anthropological research study on health-seeking behaviours was undertaken by a humanitarian non-governmental organization (NGO) in a rural village in an African country where the prevalence of child global acute malnutrition was high. The study focused on health-seeking practices during diarrhoea episodes among children under the age of five, as diarrhoea is one of the underlying causes of child undernutrition. The research aimed to study access to and utilization of health services. The country's national ethics review committee approved the research.

Qualitative fieldwork was conducted which aimed to better understand the cultural values and practices related to the therapeutic path of children with diarrhoea. Interviews were conducted with parents and other key informants in the village (e.g. community leaders, elders, traditional healers).

Consent forms were signed by the participants, but as the NGO was mostly known in the area as an assistance provider, it was not always clear to the researchers whether participants freely consented to take part in the research or whether they assumed they had to participate in order to receive assistance, or out of gratitude.

During data collection, the investigator found that a traditional treatment for diarrhoea among baby girls (from three months of age) was female genital mutilation (FGM) . This practice was intended to remove "impurity" that interfered with a girl's well-being. FGM was practised in the village by a traditional healer with a razor blade and without hygienic precautions. "If the diarrhoea is caused by a worm, we have to remove the impure part of a girl's body; it will kill the worm and cure the girl," a traditional FGM practitioner said during an interview.

According to the testimonies gathered during the research, FGM is highly valued in the local culture. In addition to being considered an effective traditional cure for girls' diarrhoea, FGM is part of the accepted and expected identity of a woman. "Uncircumcised" girls are marginalized, are a source of shame for their family and have difficulty finding a husband. FGM also has religious and social significance. This act is symbolically seen as a ritual of incorporation of the girl into the rest of the community.

At the global level, FGM is considered a violation of human rights, and it is also prohibited by law in the country where the research took place. "Female genital mutilation and cutting is a violation of the basic rights of women and girls," said Carol Bellamy, then executive director of the UN's Children's Fund (UNICEF), on the International Day of Zero Tolerance for FGM in 2005. "It is a dangerous and irreversible procedure that negatively impacts the general health, child bearing capabilities and educational opportunities of girls and women."

In the research setting of this case study, most of the participants in the interviews had never been to primary school and were illiterate. For them, local habits and regulations took precedence over national or international laws and codes of conduct.

The national ethical review committee[1] and the research team did not anticipate this finding, as their members did not have a deep understanding of the local culture and the norms of the specific community and individuals. Because this traditional cure for diarrhoea was an unexpected finding, participants had not been previously informed by the researchers of what they could be exposed to while they proudly exhibited their traditional culture.

When a researcher from an NGO witnesses a human rights abuse, there is always a risk of that organization, when managing the resulting conflicts, being accused of complicity, and/or violating the interests of both the individuals and international ethical standards. In this case the researcher acted in accordance with his own model of norms and values, and one based on national and international codes of law and ethics, rather than with the way in which the causal model of illness was understood locally, and the implications of this for the social construction of female identity. The researcher and the NGO decided to report the practices in a public report in order to protect baby girls from a recognized and illegal human rights abuse.

---

[1]The committee did not include lay members or representatives of the targeted communities.

However, this approach had serious consequences: it offended participants and the wider community, and led to the social rejection of girls who had not received FGM − they were stigmatized in the community − and intensified community tensions. It also jeopardized the NGO's capacity to operate in the area.

Communities felt betrayed by the NGO, as they were expecting humanitarian relief from the organization. They felt that the research was not responsive to their needs as they did not feel any benefit. On the contrary, its findings had exposed vulnerable communities and respondents to retribution from a coercive government, and endangered the local social structure.

Lessons Learned

This case study highlights the risks of exploitation of participants when researchers face conflicts of conscience and have to choose between abusing the trust of the community and protecting vulnerable individuals from violations of their fundamental rights in accordance with national and/or international laws and ethical codes. For the NGO involved, a lesson learned was that researchers need to anticipate the identification of potential ethical challenges by assessing the risks and benefits for potential participants with "due diligence" before a project commences. Risk assessments should not be a vertical and unilateral process, but rather a participatory exercise. This can facilitate the understanding of the context, as interpretations of benefits, risks and harm are specific to each setting.

In this context, it is important to engage in mediation with all stakeholders, which may result in an agreement according to which no actor needs to disown his/her values. The research could be ethically acceptable to all if the entire process and all the consequences are favourable (or at least neutral) for everyone. It is worth emphasizing that when opposing values are involved, it is crucial to engage in a discussion before taking action in order to reach an agreement. If no agreement can be reached before the research is commenced, then it is simply not possible to undertake the research involving that community, as some value gaps have proved impossible to overcome.

The NGO also learned that when the same organization is both conducting research and delivering aid in an area, biases can affect the voluntary informed consent of vulnerable participants, as well as the research design, data collection and interpretation, or the reporting of results. While power differences may be difficult or impossible to eliminate completely, steps can be taken to identify and minimize the most serious potential sources of bias, as long as thorough, transparent and culturally appropriate information has been given to participants.

Recommendations

- Carry out a thorough risk and benefit assessment involving community and participant representatives. Ethical approval should also be sought from the community, and community representatives should participate in the formal ethical review committee process.
- Beyond simply being asked for informed consent, communities should be trained and involved in the ethical approval process. Participants should be made aware of the limits of confidentiality and any duties the researchers have to report certain findings.
- Ensure effective ongoing communication (including with representatives of vulnerable subgroups). Communication mechanisms should not be dismantled after the departure of the research team from the data collection site, but must be maintained by local partners of the international researchers.
- Monitor and evaluate the process through which consent is negotiated with the community and obtained from participants.
- Participation in research should not be linked to receiving assistance, and researchers should make this very clear to participants to avoid any misunderstandings. In other words, if an assistance NGO operates in an area, it should be made clear that the benefits of assistance will be open to all, regardless of who, if anyone, works with the NGO on research.
- Further work is needed on how to approach unexpected findings that lead to fundamental conflicts of conscience for researchers. Data collection itself should be neutral. There should be a protocol in place regarding the consideration of and response to any unexpected findings.

Reference

WMA (2013) WMA Declaration of Helsinki: Ethical principles for medical research involving human subjects. World Medical Association. http://jamanetwork.com/journals/jama/fullarticle/1760318

Author Biographies

Gwenaëlle Luc is a social anthropologist working for Action Contre la Faim as the ethics focal point in the research department. She has undertaken research for international NGOs in the Democratic Republic of Congo, Chad, Kenya, Niger, northern Cameroon, and Burkina Faso.

Chiara Altare is senior research adviser at Action Contre la Faim, Paris. She previously worked at the Centre for Research on the Epidemiology of Disasters, Université catholique de Louvain, where she investigated the impact of conflict and natural disasters on the health of populations.

Chapter 3
International Genomics Research Involving the San People

Roger Chennells and Andries Steenkamp

Abstract In 2010 an international genomic research project entitled "Complete Khoisan and Bantu genomes from southern Africa" was published in *Nature* amidst wide publicity (Schuster et al 2010). The research aimed to examine the genetic structure of "indigenous hunter-gatherer peoples" selected from Namibia, and to compare the results with "Bantu from southern Africa" , including Nobel peace prize winner Archbishop Desmond Tutu. Four San individuals, the eldest in their respective communities, were chosen for genome sequencing, and the published article analysed many aspects of the correlations, differences and relationships found in the single-nucleotide polymorphisms (SNPs) (A single-nucleotide polymorphism is a variation in a single nucleotide that occurs at a specific position in a genome, where each variation is present to some appreciable degree within a population) within the sequenced genomes. A supplementary document published with the paper contained numerous conclusions and details that the San regarded as private, pejorative, discriminatory and inappropriate. The San leadership met with the authors in Namibia soon after publication, asking why they as leaders had not been approached for permission in advance, and enquiring about the informed consent process. The authors refused to provide details about the informed consent process, apart from stating that they had received video-recorded consents in each case (Hayes 2011). They defended their denial of the right of the San leadership to further information on the grounds that the research project had been fully approved by ethics committees/institutional review boards in three countries, (names of committees given to editors of this book) and that they had complied with all the relevant requirements. The San leadership wrote to *Nature*, expressing their anger at the inherent insult and lack of respect displayed by the process (Ngakaeaja 2011b). This case study details the most serious aspects of the perceived exploitative nature of the research, and the San response.

R. Chennells (✉) · A. Steenkamp
44 Alexander Street, Stellenbosch 7600, South Africa
e-mail: scarlin@iafrica.com

© The Author(s) 2018
D. Schroeder et al. (eds.), *Ethics Dumping*, SpringerBriefs in Research and Innovation Governance, https://doi.org/10.1007/978-3-319-64731-9_3

15

Keywords Genomics · San · Southern Africa · Indigenous peoples
Informed consent · Vulnerable population · Ethics committee
Institutional review board

Area of Risk of Exploitation

This case study is about the conducting of genomic research on a vulnerable
population, and it focuses on the enhanced need for respectful and authentic prior
informed consent. While the research itself is undoubtedly of potential benefit to
humankind as well as the participant population, the particular risk of exploitation
lies in the fact that certain types of information gleaned from genomic research are
essentially of a sensitive and private nature, and their publication can result in
potential embarrassment, discrimination and collective psychological damage. The
informed consent allegedly gained for this complex research project from the
illiterate San participants was never disclosed to the San leadership, and, as is made
clear below, the nature and content of the research publication was indeed dam-
aging to the community on various levels.

Specific Case and Analysis

The general population of San peoples of southern Africa is known to carry the
oldest human DNA on earth, and is consequently much sought after for
population-wide genomic research aimed at understanding aspects of human evolution. The San peoples, known to be the earliest "hunter-gatherer" populations of
southern Africa, number an estimated 100,000 individuals spread across at least five
countries, with the largest populations in Namibia, Botswana and South Africa.
Since 1986 the seven dominant linguistic groups have formed elected organizations
in each country aimed at representing and protecting the rights of their illiterate
rural populations. One of the most important roles of the San councils of Namibia,
Botswana and South Africa is to protect their people from unwanted, inappropriate
or exploitative research.

The stated purpose of the genomic research project under discussion was to
sequence the genomes of four selected San individuals, and to "characterise the
extent of whole-genome and exome diversity amongst them" – that is, the four San
and a man of Bantu extraction. In addition it set out to "compare the described
variants to known data-bases" in order to pinpoint genetic variations in
genome-wide data, and to "facilitate inclusion of southern Africans in medical
research efforts" (Schuster et al 2010).

In about 2009 researchers associated with the three universities began the process of obtaining informed consent and taking DNA samples from four selected San
elders from three linguistic groupings, described as Tuu,!Kung and Ju/'hoansi. How

the researchers communicated the methodology, aims and objectives of the complex research project via translators to the four illiterate elders will perhaps never be known: the San leadership later formally requested access to this information, but were refused. According to the published research, "all participants consented ... via video-recorded verbal consent (Bushmen)". In February 2010 the research was published – to wide publicity in the popular media – in an academic paper entitled, "Complete Khoisan and Bantu genomes from southern Africa", which was accompanied by a document containing supplementary information (Schuster et al 2010).

The acting regional coordinator of the Working Group of Indigenous Minorities in Southern Africa (WIMSA), Ben Begbie-Clench, approached the paper's authors requesting details of the informed consent process, as set out below. Mathambo Ngakaeaja, deputy director of WIMSA, subsequently wrote to *Nature* on 18 February 2011 objecting to the publication by Schuster et al., and describing how central the concept of prior informed consent was to all research affecting indigenous peoples. After commenting critically on the persistent refusal of the researchers to approach the official San leadership structures or engage meaningfully with them, Ngakaeaja stated that the purpose of his letter was "to draw attention to the absolute arrogance, ignorance and cultural myopia that is present here" (Ngakaeaja 2011a). He continued, "these researchers have basked in the glory of their publication whilst claiming smugly that they complied fully with the ethical requirements".

From the perspective of the San leadership, many aspects of this research study were deeply problematic, and would have been objected to if one of their organizations (e.g. WIMSA, the South African San Council or the South African San Institute) had been given an opportunity to consider the research before it began or to approve the final form of the document prior to publication.

The San leaders engaged respectfully with the researchers following publication, requesting details of the informed consent process. Despite much correspondence,[1] the authors persistently refused to acknowledge the need to consult with San leadership or to provide details of the informed consent documentation or process.

We set out below some of the San leadership's reasons for regarding the research project as exploitative.

Terminology

The use of words such as "Khoisan" and "Bushmen" and "hunter-gatherers" shows a lack of consultation with San leaders. All of these terms were freely used in the publication, but all are considered sensitive and problematic for different

---

[1]The emails concerned are in the possession of the principal author of this case study, who is a lawyer, but are not reproduced here in order to protect the privacy of personal data.

reasons. For example, the San object to being referred to collectively as "Khoisan" , a descriptive term coined by anthropologist Leonard Shutze in 1928 as a way of referring to Khoi pastoralist and San hunter-gatherer groups collectively (Schlebusch 2010). The word "Bushman", meaning "uncivilized people", is widely regarded as pejorative in certain contexts. The anthropologically loaded term "hunter-gatherer", frequently used in the paper and the supplementary information, implies a generally acknowledged low social status (Wynberg et al 2009). Consultation would have resulted in more acceptable uses of these and other terms.

Published Conclusions Far Removed from Genomic Research

Much of the discussion in the supplementary information document related to terms and concepts such as "hunter-gatherer", the low status of "hunter-gatherers", the payment of lobola and dowry, and marriage practices, for example:

> A feeling of inferiority associated with the "Bushmen" or "San" ethnic classification meant that many Bushmen women tried to uplift their status via marriage to Bantu men (Schuster et al 2010: suppl 3).

These conclusions could not have been drawn from the results of the genomic research, nor could they have been permitted by a process of informed consent to the collection of genomic data. The publication thus draws on and publishes conclusions drawn from other sources and disciplines, which would *not* have been permitted in a normal research consent process. The bad practice and injustice of publishing information that could not have been envisaged by the participants at the time of their giving consent would have been lessened had the authors returned to the communities before publication and tried to explain the far-reaching and sensitive nature of their findings. The San leadership, however, are unaware of any attempt by the researchers to return to the communities and explain the complex nature of the published conclusions.

Individual Versus Collective Consent

It is well known that indigenous, rural and illiterate people do not understand individuality and individual rights in the manner of the West, their identity being deeply collective and associated with their communities. This research project only obtained informed consent from the indigenous individuals who participated, while it is known and accepted that genomic research by its very nature speaks to collective issues. There is no shortage of published research ethics guidelines (e.g. NHMRC 2003, CIHR et al 2014) that set out absolute requirements for research on indigenous peoples, one of which is that collective "permission" should be obtained from the leadership, in addition to normal informed consent obtained from individuals. Not to do so is perceived as an expression of lack of respect for the community. However, one of the authors wrote to WIMSA saying, "As we are dealing with individuals in a personal manner (via their DNA) the individual has a right to participate or not as the information contained is of direct impact to that person" (Hayes 2011). This response does not take into account that genetic information also has a direct impact on family members of the participant.

Lack of Respect for or Reference to Indigenous Research Protocols

The need for "respect" to be shown to the particular community is perhaps the most important fundamental element in the indigenous research ethics guidelines referred to above. The requirement takes many forms, but can be summarized as authentic communication with the community leadership from the inception to the conclusion of the research project. None of the established suggested methods for showing respect to communities were employed in this case. The authors refused to consult with the leadership afterwards, relying upon the fact that allegedly none of the elderly and illiterate San participants had demanded to be represented by the San leadership. For that reason, they concluded that the San leadership had no say in the matter (Hayes 2011). This reliance on individual consent by an illiterate person who could have no idea of how the implications of genomic research related to the collective was and is regarded by the San (an abuse of power).

Failure of Research Ethics Committees/Institutional Review Boards

The researchers defended their methodology regarding consent and other aspects of the process by repeating that the project had been approved by no fewer than four separate research ethics committees. Yet not one of these committees referred to the published research guidelines on indigenous populations, which were readily available and with which they ought to have been familiar, despite the fact that the very purpose of the research was to examine the most famous of indigenous "hunter-gatherer" communities. In the words of Prof. Vanessa Hayes, geneticist and co-author of the *Nature* paper, these committees were formally designed to "approve, monitor and review biomedical and behavioural research involving humans with the aim to protect the rights and welfare of the research subjects" (Hayes 2011). In addition she stated that it was their duty to respect the "culture, dignity and wishes of subjects". It is the San view that they failed dismally in this duty.

Breaches of Privacy in the Findings

The paper and its supplementary information included a number of discussions and conclusions that contained intimate, personal or pejorative information. The following are some examples discussed in the context of "Bushmen-specific phenotypes" (Schuster et al 2010: suppl 8): namely, how different genetic and environmental influences come together to create an organism's physical appearance and behaviour.

1. *"Hunter-gatherer" associated with low social status*: Commentary in the paper on "traditional life-style" included the following, which contains far-reaching and unsupported assumptions:

   A feeling of inferiority associated with the "Bushmen" or "San" ethnic classification meant that many Bushmen women tried to uplift their status via marriage to Bantu men (Schuster et al 2010: suppl. 3).

2. *Lactase persistence*: The following conclusion was drawn:

   As expected for a foraging society, we found the Bushmen in our study all to be homozygous for the C-allele, suggesting an inability to tolerate milk consumption as adults (Schuster et al 2010: suppl 4).

3. *Human pigmentation*: Conclusions were drawn about levels of San melanin pigmentation, their susceptibility as a group to skin cancer, and their consequent selective advantage for survival in the Kalahari desert (Schuster et al 2010: suppl 5).
4. *Lipid metabolism and bitter taste alleles*: Complex conclusions were drawn relating to Bushmen digestive tracts, and also the ability to sense a bitter taste, a trait which would potentially assist human survival in the wilds. The "taste receptor gene" was also discussed in the context of human evolution from Neanderthal to the present (Schuster et al 2010: suppl 7).
5. *Genes related to hearing*: Drawing on the findings, the paper indulged in speculation that "Bushmen have better hearing than Europeans" (Schuster et al 2010: suppl 8).

Lessons Learned

The San leaders see the Schuster case as a telling example of the harm and disrespect that research can bring about, notwithstanding approval by ethics committees/institutional review bodies. It also highlights the need for San themselves to create their own protection mechanisms.

With this in mind, the San held a consultative workshop in September 2014 comprising San leaders from Botswana, Namibia and South Africa, as well as genomic researchers, ethicists and lawyers. The purpose of the workshop was to discuss the San's perception of the exploitation inherent in the approach followed by the Schuster research, and to propose a San response to ensure that such research could never take place again.

In 2016 the San held two further workshops under the auspices of the TRUST project[2] designed to take the earlier discussions further and to consolidate proposals aimed at ensuring that the San could in future manage and control research being done on their communities. These proposals are set out below.

Recommendations

The following recommendations emerged from the San workshop aimed at preventing exploitation in research.

- Collective permission must be obtained for all research to be carried out on San individuals or communities.
- The San Council is the elected organization in South Africa mandated to engage in this process with researchers.
- The San have since developed a San Code of Research Ethics (San Council 2017) that has to be completed by all prospective researchers. This code contains a number of requirements relating to the need for research to be both respectful and useful to the San peoples, including:
  - early identification of research useful to the San
  - joint development, where appropriate, of design, content and methodology of all aspects of the research
  - full details provided in advance of all aspects of the research, including (potential) benefits to the San
  - commitment to pre-publication consultation, where appropriate, and post-publication feedback to the community

References

CIHR, NSERCC, SSHRCC (2014) Chapter 9: Research involving the First Nations, Inuit and Métis peoples of Canada. In: Tri-council policy statement: ethical conduct for research involving humans. Secretariat on Responsible Conduct of Research, Ottawa ON. On behalf of Canadian Institutes of Health Research, Natural Sciences and Engineering Research Council of Canada, Social Sciences and Humanities Research Council of Canada. http://www.pre.ethics.gc.ca/pdf/eng/tcps2-2014/TCPS_2_FINAL_Web.pdf

Hayes V (2011) Personal email communication to B Begbie-Clench, WIMSA, 11 May

Ngakaeaja M (2011a) Letter to editor of Nature, 18 February

Ngakaeaja M (2011b): Personal email communication to B Begbie-Clench, WIMSA, 11 May

---

[2]TRUST is a European Union project with the main goal of catalysing a global collaborative effort to improve adherence to high ethical standards in research around the world. http://trust-project.eu/

.

NHMRC (2003) Values and ethics: guidelines for ethical conduct in Aboriginal and Torres Strait Islander health research. National Health and Medical Research Council, Australia. https://www.nhmrc.gov.au/_files_nhmrc/publications/attachments/e52.pdf

San Council (2017) San Code of Research Ethics. http://trust-project.eu/san-council-launches-san-code-of-ethics/

Schlebusch C (2010). Issues raised by use of ethnic-group names in genome study. Nature 464:487. http://dx.doi.org/10.1038/464487a

Schuster SC, Miller W, Ratan A, Tomsho LP, Giardine B, Kasson LR, Harris RS, Petersen DC, Zhao F, Qi J, Alkan C, Kidd JM, Sun Y, Drautz DI, Bouffard P, Muzny DM, Reid JG, Nazareth LV, Wang Q, Burhans R, Riemer C, Wittekindt NE, Moorjani P, Tindall EA, Danko CG, Teo WS, Buboltz AM, Zhang Z, Ma Q, Oosthuysen A, Steenkamp AW, Oostuisen H, Venter P, Gajewski J, Zhang Y, Pugh BF, Makova KD, Nekrutenko A, Mardis ER, Patterson N, Pringle TH, Chiaromonte F, Mullikin JC, Eichler EE, Hardison RC, Gibbs RA, Harkins TT, Hayes VM (2010) Complete Khoisan and Bantu genomes from southern Africa. Nature 463:943 − 947. http://dx.doi.org/10.1038/nature08795

Wynberg R, Schroeder D, Chennells R (2009) Indigenous Peoples, Consent and Benefit Sharing. Springer, Dordrecht

Author Biographies

Roger Chennells works as legal adviser to the South African San Institute (SASI) and has been senior partner in the human rights law practice Chennells Albertyn, Stellenbosch, since 1980. Specializing in labour, land, environmental and human rights law, he has also worked for Aboriginal people in Australia.

Andries Steenkamp who died in 2016, was a South African community leader and the long-standing chairperson of the South African San Council. As a representative of the San peoples he was involved in many research projects, where his primary task was to understand the mutual needs in the research relationship, and to ensure that his community both provided and received benefits.

Chapter 4
Sex Workers Involved in HIV/AIDS Research

Anthony Tukai

This case study shows that equitable relationships between researchers and research participants are about a lot more than informed consent. For instance, the higher level of research literacy among the sex workers that was achieved in the Majengo clinic is a model for others to follow.

Doris Schroeder.

Abstract This case study is written as a personal story by an outside support worker starting to engage with sex workers, a vulnerable and stigmatized population in a Nairobi slum. We hope the shared experiences will give better insight into the difficulties faced by members of this key population as they eke out a living. It is also a positive case study, not one of exploitation, despite sex work being illegal in Kenya.

Keywords Clinical trials · Sex workers · Kenya · Women · Empowerment

My Experience Visiting Majengo

I took an assignment with the Sex Workers Outreach Programme (SWOP), a leading sex workers' health organization in Kenya that promotes the health, safety and wellbeing of sex workers, as well as affirming their rights as workers and as people. The programme is funded by CDC-PEPFAR[1] through the University of Manitoba, Canada. I began my assignment by visiting the Majengo slum where SWOP runs a health clinic targeting sex workers living in and working from these informal settlements.

I am Kenyan with a background in social work and public health. My public health interest is in HIV prevention, while my social work interest is in

---

[1]The US President's Emergency Plan for AIDS Relief, as implemented by the Centers for Disease Control and Prevention in the USA.

A. Tukai (✉)
Sex Workers Outreach Programme (SWOP), P. O. Box 30709, Nairobi 00100, Kenya
e-mail: tonytukai@gmail.com

© The Author(s) 2018
D. Schroeder et al. (eds.), *Ethics Dumping*, SpringerBriefs in Research and Innovation Governance, https://doi.org/10.1007/978-3-319-64731-9_4

interventions. My hope is to build a strong foundation to improve the health and well-being of vulnerable and stigmatized communities such as LGBTs (lesbian, gay, bisexual and transgender persons) and sex workers.

I have lived in Nairobi for the greater part of my life, but like most Kenyans I had never visited a slum in Kenya. I was prepared for the unexpected, but it was like going to a different world. What struck me first were the overcrowding and the variety of activities that the residents engaged in for survival. The area is densely populated; it felt like being in a city within a city. The road we tried to drive along to access the slum was full of people selling second-hand shoes, clothes and household items. We had to stop and wait for close to ten minutes for the hawkers to clear a path – like parting the Red Sea – so that we could get into the clinic compound. I took a walk with a health worker from the Majengo sex workers clinic to meet some of the sex workers who live and work in the area. We saw dirty alleys, open sewers and lots of trash. There were women doing laundry on the sidewalks, and some sitting beside their doorsteps. Men were going in and out of the houses or just walking around, many looking as if they had been drinking heavily. I saw hardly any children; those I did come across were playing outside unsupervised. The more fortunate children were presumably attending school in other parts of the city, while many of the rest were at the dumpsites, trying to earn money by scavenging for recycling companies. The houses were small and squeezed together, poorly built with rusted metal roofs.

The sex workers to whom I was introduced on the narrow pathways and by the doorsteps were friendly, saying "Hello!" and "Karibu!" (welcome). One of them ushered us into her tiny room. She had been doing sex work since she was a teenager and now looked to be in her mid-50s. She was skinny; I think she weighed no more than 40 kg. Her single room was small and cramped, with no space for a kitchen area or a living room. But two beds were squeezed in. One of the beds, she said, was her "office ... where I service my clients, and the other one is where I sleep when not working". Hanging on top of her bed was an assortment of underwear in different styles and colours. She smiled and said, "Some of my clients prefer me to wear different colours and shapes of underwear, so I keep this for them."

Seeing the Majengo slums and experiencing something of the life there was an eye-opening experience that I will not forget for the rest of my life.

About Majengo

The Majengo slums are about three kilometres away from Nairobi's central business district. One of the oldest slums in the country, it is located between Gikomba market (the biggest mitumba, or second-hand clothes market, in East Africa) and Eastleigh, a commercial hub that is now known as Little Mogadishu due to the huge number of Somali immigrants living in the area. Majengo can be traced back to colonial times in the 1920s, when it was occupied by East African railway builders and those serving them.

In her book *The Comforts of Home: Prostitution in Colonial Nairobi*, White (1990) describes how cattle epidemics, locusts, famine and drought swept through Kenya in the 19th century. A lack of food and the spread of disease, including smallpox, in central Kenya caused the death of an estimated 70% of the population.

After the famine, the Nairobi economy began to boom in the mid-1920s, with men and women from neighbouring districts arriving to sell agricultural products. Many ended up staying in Majengo. Sex workers became Kenyan's "urban pioneers", and were among the first residents to live in Nairobi year-round. They frequently came from strong families (White 1990:9). Many were able to send money home to bolster rural family incomes, which were racked by upheavals. Prostitution emerged as an identifiable category of women's work, taking three forms:

- Watembezi prostitutes (from the Swahili word *kutembea*, "to walk") offered brief sexual services along the streets.
- Malaya (the term means "prostitutes" in Swahili) offered more prolonged indoor domestic and sexual services.
- Wazi wazi ("open") prostitutes sat in front of their houses, calling out their prices raucously and aggressively.

For some women, sex work was casual and intermittent: "He was hungry for sex and I was hungry for money" (White 1990:85). For others, it was the only way to survive: "[W]e were hungry, we had to go with men to get money, or have no money" (White 1990:79).

Majengo, also known as Sofia Town, was once an entertainment spot for British soldiers who frequented the village to watch cultural performances by mostly female groups. During the colonial era, Majengo grew into quite a popular area, but without the provision of adequate shelter. Its population today is estimated to be more than 150,000 people of all ages and different ethnicities. It is divided into the four smaller settlements of Sofia, Mashimoni, Kitanga and Digo. The women continue to sell sex, filling a gap for men whose wives, girlfriends and families remain back home in rural Kenya. In addition, men from other countries continue to visit Majengo for sex.

The Majengo Clinic

The Majengo Clinic is a medical facility for low-income and medically underserved communities. Within a larger compound there is a special clinic, also known as the Special Treatment Centre (STC), that has offered sex workers a safe space since the mid-1980s. For a long time it was the only public health centre in Kamukunji, Nairobi, providing sex workers and their clients with treatment for sexually transmitted infections (STIs). With funding from the Canadian government and the assistance of the public health authority of Nairobi City Council, researchers from the universities of Oxford, Nairobi and Manitoba worked to improve existing resources and provide basic outpatient medical services to the Majengo community of female sex workers. In the mid-1980s, the World Health Organization (WHO) designated their operations as a WHO collaborating centre for sexually transmitted diseases (STDs). Among the common ailments treated were classic STDs, malaria and typhoid. Currently, the clinic offers comprehensive HIV prevention and treatment services, birth control methods, gynaecological examinations, and TB tests and treatment, in addition to supporting the management of assorted HIV/AIDS-related opportunistic infections. It also serves as a research facility for the collaborating researchers, who run two HIV-integrated activities: HIV research and HIV care and treatment. More than 5,000 sex workers receive care at the clinic, 3,200 of them enrolled in clinical research studies.

Majengo Research

The Majengo Observational Cohort Study (MOCS) started in the late 1980s, and is a long-term cohort study of disadvantaged female sex workers in Nairobi. The study, as expected, has contributed to the development of several candidate vaccines against HIV.

HIV research studies started when Dr Frank Plummer, a Canadian scientist who was the principal investigator undertaking research on STIs in Majengo, discovered that about two-thirds of women visiting the clinic had tested positive for the virus in 1985. This changed the focus of his research from general STIs to include the epidemiology of HIV in Africa.

Plummer and his team later discovered that a small number of the women had apparently developed immunity to the HIV virus despite long-term exposure through sex with infected clients. This led to other studies aimed at understanding the epidemiology and immunobiology of HIV and the risk factors associated with its spread. Blood, cervical, vaginal and saliva samples were drawn from women in this cohort, with their consent. One of the key findings was that when some of the "HIV-resistant" women took breaks from sex work – for example, to visit family or pursue alternative employment – temporarily stopping their exposure to HIV, they rapidly lost their immunity and became significantly at risk of HIV-infection on resuming sex work.

Majengo Research Participants

"Prostitutes" is what they called them in the past. Then they were known as commercial sex workers, and now the term is sex workers. "I don't know what they will call them next," said one of the Majengo clinic workers during my visit. "Sex worker" is the term used by researchers and policymakers and includes female, male and transgender adults aged over 18 years who sell consensual sexual services in return for cash or payment in kind, and who may sell sex formally or informally, regularly or occasionally. It's a word used by people who think the word "prostitute" is impolite or offensive.

Sex work is classified under Kenya's Penal Code as illegal (Laws of Kenya 2014), and it entails a stiff penalty. It is seen as an "immoral activity" rather than a form of labour, and many believe that sex workers deserve to be punished.

At first sex workers were nervous to register with the SWOP clinic, because they feared that their personal information would be shared with the Kenyan law enforcement agencies. Once they were assured that the information gathered through unique identifiers and biometric tools was for research purposes and would not be shared with any third party, they registered in droves. They also signed informed consent documents for the different research studies undertaken. In return, the SWOP team provided and continues to offer free health care including HIV management. One of the reasons why the clinic has a good record on research ethics is its engagement work with the community that is involved in the research.

Research Literacy Among the Sex Workers

The research has built up long-term relationships between the researchers and the women sex workers through peer leaders and educators who engage in dialogue and negotiations with the scientific investigators about the terms and conditions for participation in the research. Over time, these activities have helped to develop and formalize a "community" among the sex workers that did not previously exist. The partnership has enabled a wide range of benefits in the research cohort and wider community, such as health education, free distribution of condoms, and the provision of free treatment for a range of STIs. In addition, it has led to effective referral for other health care requirements, such as non-communicable diseases, cancers and surgical procedures including hysterectomy. Such services would probably not otherwise have been available to these women.

These peer educators are themselves sex workers. They educate women about their rights, promote behavioural change, distribute condoms and provide referrals to health clinics. Peer educators also address workers' concerns, whether about personal issues, services offered, or the research they are a part of. They are the gatekeepers of the sex workers' community.

Education about condom use has given sex workers the confidence to negotiate this with their clients. Over time, 100% condom use has been achieved with casual clients, but regular clients still remain a challenge. Peer educators have also been active in the provision of general information on the research consenting process. Capacity building on the consent procedures undertaken over the years by the SWOP team seems to have borne fruit. Sex workers currently involved in the pre-exposure prophylaxis (PrEP) studies[2] have stated that they are not subject to any pressure in deciding whether to participate in any of the research projects. "We are free to refuse to consent to any research, be it from SWOP or any other," stated one of the participants. "Consent is voluntary and has always been voluntary at the Majengo clinic," said a sex worker who was also a peer educator.

Ethical Concerns and Benefits

The sex workers have long been collaborating with researchers from Kenya, South Africa, Europe and Canada. The Majengo clinic has also been providing better health care than is offered at other public health facilities. There are obvious issues around informed consent and the possible exploitation of the sex workers in the studies that constantly have to be dealt with. For example, do the sex workers really understand what they are consenting to, or do they trade participation for access to better and free health care? The peer educators, in my judgement, are influential. Do they therefore play a big role in the willingness of sex workers to participate in the studies? Does the collective opinion of the sex worker community on particular studies have a major influence on individual willingness to participate, thus diluting autonomy and self-determination?

These issues have been raised before – for example, in a newspaper article headed "Sex slaves for science?" (Nolen 2006) and in *Benefit Sharing: From Biodiversity to Human Genetics* (Schroeder and Cook Lucas 2013) – and they demand answers. In addition, Andanda and Cook Lucas (2007:9) have stated that:

> In the Majengo case, the original, routine issues of negotiation and decision-making related to the conduct of the research studies only involved researchers and administrators from the relevant universities and institutions. ... There was no formal inclusion of representatives from the sex workers in any of these negotiations.

While writing this case study, I asked one of the peer educators about the inclusion of sex workers in decision-making in the past. She confirmed that inclusion and genuine partnership had not been emphasized previously, but she added: "Now we are enlightened, this would not happen at the moment without our consent. We must be part of the decision-making". The long-term engagement of the clinic with research participants in the spirit of ethical research has therefore,

---

[2]Pre-exposure prophylaxis, or PrEP, is a way for people who do not have HIV but who are at substantial risk of getting it to *prevent* HIV infection by taking a pill every day.

over time, led to improvements in the positioning and negotiation skills among the peer leaders/educators. This is easily noticeable on the ground. Other factors noted include:

- The women's health has improved because of their access to education and high-quality care, which has reduced HIV incidence, the disease burden and mortality.
- Important findings about HIV infections are shared with the sex workers' community as they emerge. This has a great impact on the health of sex workers generally, since both partners now practise evidence-based interventions and programming. This will become even more prevalent as a greater understanding of novel prevention strategies emerges, and as these strategies are adopted.
- Sex workers involved in the studies have increased their self-worth and agency by becoming valued partners in the research and by developing a sense of community among themselves. It is important not to romanticize this, because the women's lives are fraught with difficulty, but it has to be noted that sex workers have been able to counter assaults on their self-worth due to the illegality of sex work in Kenya by developing a new emphasis on their rights. Bandewar et al. (2010) argue that participation in the MOCS has improved and enriched sex workers' lives, because community engagement activities have helped create a community that did not exist independently. Majengo sex workers – as part of the growing sex workers' movement in Kenya – have formed an association called the Kenya Sex Workers Alliance (KESWA). This is a local chapter of the global sex worker alliance, whose mandate is to train sex workers about their human rights. "Sex work is work!" is an everyday slogan among the Nairobi sex workers.

Poor enrolment in the ongoing PrEP demonstration project, despite a huge number of potential at-risk HIV-negative participants from the cohort, presents some real food for thought. In my discussions with the sex workers' representatives, they pointed out that community education, demand creation and advocacy for PrEP among the sex workers were done poorly. The researchers and policymakers had not fully engaged the community in promoting the project. Therefore uptake of the novel intervention, despite its potential, will remain poor so long as the sex workers' community is not educated and involved in the grass-roots advocacy processes. Inclusion and the community buy-in and support are crucial to progress. This finding also confirms that the Majengo sex workers do indeed practice self-determination in the consenting process.

Conclusion and Looking Forward

At a recent TRUST[3]-sponsored high-level meeting in Nairobi, the peer educators' demands for inclusion went a notch higher. They insisted on being part of the ethics board that approved any research study involving sex workers. They also asked to be included in the technical working group to advise on issues concerning sex workers.

The ethics concerns for a group of sex workers from the Nairobi slums are obvious. I would like to end with two observations. First, to be considered vulnerable in a research context does not mean to be weak or to need others always to speak for one. Many of the sex workers I have met are very clear when expressing their concerns and suggesting ways forward. Second and very important, sex workers have increased their self-worth by participating in past and ongoing studies. They are now more empowered to make their own choices, whether these choices concern the way they receive their health services from SWOP or their decisions about participating in research projects.

References

Andanda Pamela, Cook Lucas J (2007) Majengo HIV/AIDS research case: a report for GenBenefit http://www.uclan.ac.uk/research/explore/projects/assets/cpe_genbenefit_nairobi_case.pdf

Bandewar S, Kimani J, and Lavery J (2010) The origins of a research community in Majengo observational cohort study, Nairobi, Kenya. BioMed Central Public Health 2010. doi:10.1186/1471-2458-10-630

Laws of Kenya (2014) Penal Code (Cap. 63) 153–156. Government Printer, Nairobi, Kenya. http://www.kenyalaw.org/lex//actview.xql?actid=CAP.%2063

Nolen S (2006) Sex slaves for science? The Globe and Mail (Canada), 7 January. http://www.theglobeandmail.com/life/sex-slaves-for-science/article20407422/

Schroeder D, Cook Lucas J (2013) Benefit sharing: from biodiversity to human genetics. Springer, Dordrecht

White L (1990) The comforts of home: prostitution in colonial Nairobi. University of Chicago Press, Chicago IL

---

[3]TRUST is a European Union project with the main goal of catalysing a global collaborative effort to improve adherence to high ethical standards in research around the world. http://trust-project.eu/.

Author Biography

Anthony Tukai is a behavioural scientist with an interest in HIV prevention and interventions among vulnerable and stigmatized groups. Anthony has worked with several non-profit organizations in the UK and the US. He currently lives in Kenya, working to improve the health and well-being of lesbian, gay, bisexual and transgender communities and sex workers. Anthony works in Nairobi for Partners for Health and Development in Africa.

Chapter 5
Cervical Cancer Screening in India

Sandhya Srinivasan, Veena Johari and Amar Jesani

Abstract Three clinical trials took place in India between 1998 and 2015 in urban and rural areas of Mumbai, Osmanabad and Dindigul. The trials aimed to determine whether trained health care workers could conduct cervical cancer screening in a community using cheap methods of testing – primarily visual inspection with acetic acid – to reduce the incidence and mortality rate of cervical cancer. The clinical trials were conducted on approximately 374,000 women, of whom about 141,000 were placed in the control arm (no screening). Although the standard of care for testing of the disease in India has been cytology screening (or Pap smear) since the 1970s, screening for cervical cancer was not available universally under a government programme, and for the study purposes the standard of care was therefore misconstrued to be no screening. Known and effective methods of screening for cervical cancer were therefore withheld from 141,000 women in areas where it was known to be of high incidence and prevalence. This placed them at a known risk of developing invasive cervical cancer, and dying from it, because it was not detected and treated in time. Two hundred and fifty-four women in the no-screening arm died due to cervical cancer as per the latest published reports on the three trials. A no-screening control arm would not have been allowed in the USA, but was accepted by the US funders for clinical trials in India. It is imperative that ethical standards for research be applied equally across nations to prevent "ethics dumping" and protect the rights of human research participants in research, no matter where they are located on the globe.

Keywords Clinical trials · India · Cervical cancer · Women · Standard of care

The date periods for deaths in the no-screening arms are taken from the dates quoted in the last available publication on each trial. They are: 98 in Mumbai 1998–2011 (Shastri et al. 2014), 64 in Osmanabad 2000–2007 (Sankaranarayanan et al. 2009), 92 in Dindigul 2000–2006 (Sankaranarayanan et al. 2007). The Mumbai trial reported findings up to 2011, though the trial would not have ended before 2015.

S. Srinivasan (✉) · V. Johari · A. Jesani
Indian Journal of Medical Ethics, 8 Seadoll, 54 Chimbai Rd, Bandra West, Mumbai 400 050, India
e-mail: sandhya_srinivasan@vsnl.com

© The Author(s) 2018   33
D. Schroeder et al. (eds.), *Ethics Dumping*, SpringerBriefs in Research and Innovation Governance, https://doi.org/10.1007/978-3-319-64731-9_5

Area of Risk of Exploitation

While the trials described in this case study showed a number of ethical short-comings, the main area of risk of exploitation was a placebo arm – no screening for cervical cancer despite high incidence and prevalence – instead of provision of an accepted standard of care.

Context

Medical and public health research had crossed national boundaries during colonial times; but controversies on ethical violations in research conducted by those from high-income countries (HICs) in low- and middle-income countries (LMICs) became a real focus once higher ethical standards were established in the HICs. LMICs however have continued to lag far behind in bringing such standards into their legal and ethical systems.

This unevenness in ethical and legal standards has been used by HICs to carry out research at reduced financial costs in LMICs. Many participants in such research have suffered avoidable injuries and deaths. The international bioethics debate has chastised researchers from HICs for practicing "double standards" (Macklin 2004), taking advantage of vulnerable people in vulnerable nations and thus "exploiting" them for their own scientific goals and profit motives. Inequities among researchers, and in ethics standards, have since become major issues of concern in international collaborative research.

The globalization of neoliberal economic policies has pressured LMICs to open their markets and deregulate their economies. The establishment of the World Trade Organization in 1995 created an international trade regime favourable to HICs. One major issue in international trade is the "dumping" of cheap and/or substandard commodities by powerful nations into the economies of less powerful nations, with a devastating negative impact on their economies (Howell and Ballantine 1998).

"Ethics dumping" follows the same pattern as dumping in trade, but in slightly different ways. Ethics dumping takes place because doing such research is either not possible at all in the HIC concerned or entails high costs due to the value attached to the ethical standards it is required to follow. This is matched in the low- or middle-income country (LMIC) by either a lack of adequate ethical standards in its guidelines or a failure to convert such guidelines into law and mandatory requirements and enforce them. At the same time, the suffering of many people from a range of communicable and non-communicable diseases may render such research relevant to the LMIC, and may also tempt local scientists to undertake it with inadequate ethical standards, in order to find well-intentioned solutions.

While research of this kind may or may not provide an early solution to a medical problem suffered by people, the need for a solution invariably tends to provide a justification for using a lower ethical standard, according less importance to respect for participants, or to the avoidable injuries and deaths of vulnerable subjects. Overall however, it causes irreparable harm to the nation's desire to bring ethical standards up to an international level.

We provide an example of ethics dumping in three trials conducted from 1998 to 2015 in urban and rural India on testing for cervical cancer. These were funded by the USA's National Institutes of Health (NIH) and the Bill and Melinda Gates Foundation (BMGF), a private foundation that supports public-private partnerships in the development of technological solutions and their inclusion in government programmes, in collaboration with the International Agency for Research on Cancer (IARC) in France, a specialized cancer agency of the World Health Organization (WHO).

These trials have been condemned as unethical by public health experts and ethicists because the participants were not offered the same level of protection and consideration as participants in HICs would have been. Women in the no-screening arm of the three trials were merely observed to determine how many would get cervical cancer and how many would die, if they were never screened. Issues relating to informed consent, the use of placebo or control arms of the trial (in this case no screening) despite awareness of and the in-principle availability of well-known effective methods of testing for cervical cancer (e.g. Pap smear), a lack of proper supervision in the intervention arm of the trial, and irreversible harm to the women participants have marred these trials and resulted in human rights violations.

Background on Cervical Cancer Screening in India

Cervical cancer is the fourth most common cancer in women worldwide, with 85% of the global burden of disease in LMICs (Ferlay et al. 2013). It is a leading cause of cancer mortality in Indian women over the age of 15, and too often women die because they do not get prompt diagnosis and treatment. Researchers note:

> Nearly 70% of cervix cancer patients in India present at stages III and IV. Around 20% of women who develop cervix cancer die within the first year of diagnosis and the 5-year survival rate is 50% (Mittra et al. 2010).

This cancer affects poor women the most, especially those living in rural areas, because they are less likely to get screened and treated, and therefore more likely to develop invasive cancer and die from it (Krishnan et al. 2013).

In HICs, regular screening programmes for the early detection of precancerous lesions, and their prompt treatment before they progress to invasive cancer, have led to a reduction in incidence of and deaths from cervical cancer (Sankaranarayanan et al. 2001). The international standard of screening is cytology, or the Pap smear, an examination of cells on the surface of the cervix for precancerous lesions. Another test involves the DNA of the human papillomavirus (HPV), a viral infection closely associated with the development of cervical cancer. The HPV test, which is manufactured by various companies, is being advocated for routine use in HICs, where it costs substantially more than cytology.

Cytology screening has been used in Indian public health services since the 1970s and is available in all major hospitals in the country. Since at least 2001 it has been advocated for inclusion in the government's cancer control programme services (Sankaranarayanan et al. 2001). In 2006, guidelines developed by the Indian government and WHO advocated the use of the Pap smear at district level, along with a cheaper, simpler screening method at the primary health centre level (National Cancer Control Programme 2006). The HPV test is available in the private sector in India, but it is very expensive. Though cytology is available all over India, researchers have held that it is not feasible for population screening in a country like India:

> Cervical cancer prevention researchers and advocates have argued that the standard approach in high-income countries, namely cytology-based screening, is difficult to establish in LMICs where laboratory infrastructure; trained personnel, such as cyto-technicians and pathologists; and continuous quality assurance processes are largely unavailable … Consequently, research has focused on evaluating screening approaches requiring less training and fewer clinic visits and using existing (or minimal additional) human resources (Krishnan et al. 2013).

An inexpensive cervical screening method is visual inspection of the cervix to detect precancerous lesions. Since at least the 1990s, studies have been conducted of various visual inspection methods, with or without magnification, and after application of contrast chemicals such as acetic acid or iodine to highlight precancerous lesions. These methods do not need to be conducted by a medical professional. By 1999, visual inspection of the cervix after application with acetic acid (VIA) was considered a "promising approach in the detection of cervical neoplasia" (Sankaranarayanan et al. 2003) for cancer prevention programmes. VIA was being advocated for inclusion in the cancer screening programme as early as 2001, but it was felt that definitive information on the value of VIA was still lacking.

Study Design

The value of a screening intervention as a public health measure is judged by various criteria: sensitivity, specificity and positive predictive value of the test; the feasibility of implementing it in a health programme, its cost-effectiveness, and its impact on incidence and mortality. Such information is gathered through various types of research, including cross-sectional studies, mathematical modelling, implementation projects and cluster randomized controlled trials (CRCTs).

Within the scientific community, the CRCT is a classic trial design to evaluate an intervention in the community. The CRCT provides the gold standard of evidence necessary for making public policy decisions. CRCTs test an intervention (preventive or therapeutic) for a disease or condition by giving it to a "cluster" of people, and comparing the results to a control group of clusters, who are given another intervention. The clusters can be slums within a municipal ward, or villages covered by a single primary health centre. The group or sample is chosen from a larger community using a system of randomization that is meant to eliminate all differences between the two groups (e.g. age or parity) other than the intervention being studied.

When there is no existing effective intervention for the disease being studied, then a trial may compare the intervention to a placebo (e.g. a "dummy pill"). When a non-drug trial tests a preventive intervention such as screening, then the "placebo" arm is a "no-screening" arm. However, ethical guidelines governing the use of placebo in research severely restrict the use of placebo or "no intervention" *if an effective treatment or test already exists for the disease being studied.* This is to ensure that research participants in the control arm do not receive a lower standard of care than is already known to be effective, and are not therefore disadvantaged by their participation in the study. This had been asserted in a number of national and international documents published prior to and during the three trials undertaken in this case study (WMA 2008; ICMR 2000, 2006; CIOMS 2002). The World Medical Association's *Declaration of Helsinki* first introduced strict guidelines on the use of a placebo control in 2000 (WMA 2000).

Three Cluster Randomized Controlled Trials of VIA with "No Screening" Controls in India

In a review of cervical cancer screening in LMICs, R. Sankaranarayanan et al. described research on cervical cancer screening in India, which included studies of the impact of awareness and health education, and cross-sectional studies of various visual inspection-based approaches as well as HPV testing. They concluded by mentioning three studies:

> There are three large, ongoing cluster-randomized intervention trials in India – in Dindigul district (Tamil Nadu), in Mumbai, and in Osmanabad district (Maharashtra) – to evaluate the effectiveness of VIA in reducing cervical cancer incidence and mortality. The intervention programme in Osmanabad district aims to address the comparative efficacy and cost-effectiveness of three different primary screening approaches in reducing the incidence and mortality: VIA, conventional cervical cytology, and HPV testing. The results of these studies are likely to provide valuable leads to the development of public health policies to control cervical cancer in developing countries (Sankaranarayanan et al. 2001).

The trials were conducted on a total of 374,000 women. The 141,000 women in the control arms of these trials received no screening for cervical cancer, but were provided with the so-called "usual care" or "standard care", consisting of health education on cervical cancer symptoms, screening and treatment, and the availability of these facilities in their localities. According to the last published report on each trial (Sankaranarayanan et al. 2007, 2009; Shastri et al. 2014), a total of 548 women were recorded to have died in the trials, 254 of them in the no-screening control arms.[1]

The use of no-screening control arms went against all established ethical principles, as articulated in national and international guidelines: namely, that new interventions must be tested against a proven effective method. In the case of the VIA trials, cytology screening was a proven effective method, and it was available in health services all over the country, including in the institutions which conducted these trials.

When a controversy about these trials using a no-screening control broke out, one of the investigators stated: "Whenever a new intervention is evaluated, it is compared with the standard of care existing in the country". In India, he wrote, there "is no organised or large-scale opportunistic cervical cancer screening programme" anywhere in the country. As a result, "[t]he standard of care for cervical cancer control in India is clinical diagnosis and treatment of invasive cancer only when symptomatic women seek medical attention" (Sankaranarayanan et al. 2011). Another researcher stated: "Pap smear cannot be considered the standard of care in India, not only because of the lack of infrastructure and trained manpower, but also because it is not cost-effective" (Pramesh et al. 2013).

All the women recruited in these trials were poor and socially disadvantaged, and thus highly vulnerable. The Mumbai study was conducted on women in slum clusters living in tenements, shanties on open ground, or makeshift huts on the pavements and along the railway lines. Osmanabad, Maharashtra, is "a predominantly rural and socio-economically backward district with a high incidence of cervical cancer" (Sankaranarayanan et al. 2005). Between 25–30% women lived in thatched roof houses. Dindigul, Tamil Nadu, is a rural district whose high incidence of cervical cancer was a reason to choose it as the site of this VIA trial. Some 65–75% of the women in Osmanabad and Dindigul and 40% in Mumbai had no formal education. The average age of the women in these trials was 40–45 years (range 30–59). They would have had poor access to health care, whether because of cost or the inconvenience of long waiting lines and ill-equipped public services or the low priority given to self-care. Though it is known that the vast majority of women, particularly after they have given birth, suffer from various gynaecological symptoms, 90% of women in the Mumbai trial had never visited a gynaecologist with their complaints.

The Mumbai Trial

The first study to start was in Mumbai, at the Tata Memorial Hospital and Centre (TMC), a national centre of excellence for cancer research and policy. The study,

---

[1]The figures are based on the start and cut-off dates given in the study reports: Dindigul: 2000–2006, Osmanabad: 2000–2007, Mumbai 1998–2011. The Mumbai trial concluded in 2015, but reported results as of 2011.

entitled "Early detection of common cancers in women in India", was funded by the US National Institutes of Health. The study initially sought to find out if repeated rounds of screening using inexpensive techniques would reduce mortality from cervical cancer.

Women community health workers educated up to the tenth grade were trained to conduct screening with VIA and also to do clinical breast examination for the detection of breast cancer. They were required to be supervised, and about 10% of women screened were also to be tested by the researchers for cross-checking of the results. Women in both arms were given health education on the causes of cancer. They were also told about the need for screening, and that the screening and treatment were available. Then the women in intervention/experimental arms were given screenings for cervical cancer, while women in the control arms were given no screening at all.

The trial started in 1998 and concluded in December 2015. A total of 75,000 women in the intervention arm and 76,000 women in the no-screening arm were recruited into this trial. Each woman was in the trial for 17 years. *Women in the intervention arm* were given health education and screening four times, i.e. once every two years. Those who tested positive were directed to TMC, where they were given confirmatory tests and treatment if needed. After the four rounds of screening were over, the women were then contacted four times, once every two years, for follow-up. *Women in the control arm,* on the other hand, received health education only once, were not offered any preventive screening for carcinoma cervix, and were observed through surveillance for 17 years. Every two years, through active surveillance, data of women in the control arm were collected to find out the number that developed cervical cancer or died as a result of it. In 17 years, seven rounds of active surveillance were carried out in both arms to document the development of cervical cancer and deaths due to it.

Changes were made to the study protocol over a period of time. The intervention was initially "direct visual inspection" without any magnification or contrast, a technique that had been judged obsolete before this trial began, and was later changed to VIA. The sample size increased from about 35,000 in each arm initially to about 75,000 in each arm. The objectives were later amended to include reduction in the incidence of cancers. These details do not appear in the published reports of the study. Cross-checking of test results by the researchers was also performed for fewer than 10% women in the intervention arm.

In 2011, an American physician filed a complaint with the US government's Office of Human Research Protections (OHRP), relating to the Mumbai and Osmanabad trials. An application for documents was also filed by a journalist under the US Freedom of Information Act. The OHRP stated that its jurisdiction was limited to trials funded by the US government and did not apply to research funded by the Bill and Melinda Gates Foundation (BMGF) , a private party (Suba 2014).

The OHRP's investigation found irregularities in the functioning of TMC's institutional review board: standard operating procedures had not been followed, meeting minutes were not documented, and decisions were taken without a quorum. The OHRP also found discrepancies in the informed consent document between the

English and the local language translation (Marathi). The English form gave information on cervical cancer, the tests required for its detection and where testing was available, but the Marathi form did not.

The OHRP did not find the no-screening arm of the trial to be unethical. By 2011, 98 women who had entered the control arm of the Mumbai trial and received no screening, only health education, had died of cervical cancer. The results of the Mumbai trial were announced at the 2013 meeting of the American Society of Clinical Oncology. The researchers announced that a test for cervical cancer, using just vinegar and conducted by trained health workers, could bring mortality from cervical cancer down by 31% (ASCO Post 2013). The findings were reported extensively in the press.

Osmanabad Trial

In October 1999, TMC with its Rural Extension Project and the Nargis Dutt Memorial Cancer Hospital started a second trial, in the Osmanabad district in rural Maharashtra. They were funded in this trial by BMGF.

This trial compared the impact of a single screening of VIA, HPV test or cytology to a no-screening control arm in a CRCT. The primary outcomes were the incidence of cervical cancer and the associated rates of death. The researchers stated in their interim report:

> Whether a screening program using VIA or HPV testing will be followed by a reduction in disease burden and the cost–effectiveness of these alternate approaches based on real program-based information remain to be established. These approaches need to be evaluated in comparison with the established standard cytological screening, with respect to their comparative efficacy and cost–effectiveness, before recommendations can be made concerning their introduction in a public health context (Sankaranarayanan et al. 2005).

Women in the intervention arm were identified through household surveys, and those who consented to be in the trial were given information on cervical cancer and its prevention, and invited to screening camps in each village where trained midwives conducted the screening. Depending on which intervention arm the village belonged in, the women received VIA, Pap smear or the DNA test for HPV. Women with positive VIA tests were given immediate follow-up tests and on-the-spot treatment if appropriate; or they were referred to the Nargis Dutt Hospital for further treatment. Samples from the cervix were taken from women in the cytology and HPV arms, and the results sent to them in two weeks. Those with positive tests were given appointments at the hospital for follow-up. Women in the control arm were given education on cancer and its prevention and information about the services available at the Nargis Dutt Hospital. "Since there is little screening for cervical cancer in India, women who did not undergo screening (control group) were considered to receive the standard of care" (Sankaranarayanan et al. 2009).

All the women were contacted just once, at the time of the intervention, after which they were surveyed and tracked through the cancer registries and death registries, until the end of the eight-year follow-up period. The trial was conducted in partnership with the IARC and the Association for Cervical Cancer Prevention (ACCP). The ACCP is a member of the IARC and both receive some funding from the BMGF. Screening was started in January 2000 and completed by April 2003 (Sankaranarayanan et al. 2005). The findings of the interim report of the Osmanabad trial ran contrary to standard wisdom:

> Our results show that a high level of participation and good-quality cytology can be achieved in low-resource settings. VIA is a useful alternative but requires careful monitoring. Detection rates obtained by HPV testing were similar to cytology, despite higher investments (Sankaranarayanan et al. 2005).

However, when the final findings were reported in 2009, the researchers concluded that while a single round of screening for HPV reduced both incidence and mortality from cervical cancer, cytology and VIA were no better than no screening at all. The researchers observed that while the test used in the trial, by Digene Corporation, was effective, a cheaper HPV test had been developed, manufactured by Qiagen, a Chinese company.

> Our results, combined with those of the Chinese study of the new HPV test, indicate that HPV testing is appropriate as a primary screening approach in low-resource settings for women who are at least 30 years of age (Sankaranarayanan et al. 2009).

These comments gain significance when one learns that in 2004, Digene had entered into a partnership with the Program for Appropriate Technologies in Health (PATH), an implementing agency for BMGF, to promote the use of HPV testing in LMICs. In 2007, Qiagen Corporation bought Digene Corporation.

Dindigul Trial

The third trial started a few months after the Osmanabad trial. In May 2000, BMGF with IARC initiated another trial of VIA, this one with the Christian Fellowship Community Health Centre hospital in Ambilikkai, Dindigul District, Tamil Nadu. The objective was to evaluate the efficacy of a single round of VIA provided by nurses, with appropriate treatment approaches, in reducing the incidence of and mortality from cervical cancer.

The women in the intervention arm were screened with VIA by trained nurses. Those found positive were offered cryotherapy on the spot, and those with larger lesions were referred for treatment. Screening was completed by April 2003. The control group received "existing care". "No active intervention was provided for the control group" (Sankaranarayanan et al. 2004). The researchers explained: "We used an unscreened control group because there are no organised screening programmes in India" (Sankaranarayanan 2007).

Information on incidence and mortality was collected from cancer and mortality registries as well as through active follow-up. Follow-up started in September 2003 and was to continue until 2012. However, by December 2006 the researchers concluded that a single round of VIA followed by appropriate treatment reduced incidence and mortality significantly. "Timely implementation of an affordable and effective screening strategy in developing countries is thus crucial, while waiting for further improvements in HPV testing, vaccine technology, costs, and its widespread use" (Sankaranarayanan 2007).

Analysis

While ethics and human rights often offer universal frameworks for research, their actual implementation differs from country to country. Basic ethical principles of research such as those of informed consent, cautions on research on vulnerable populations and the need for monitoring mechanisms to protect participants are laid out in international guidelines. The three trials described in this case study are evidence that principles of research ethics are not always translated into practice.

The VIA trials demonstrate ethics dumping, and the harm that it causes to participants in host LMICs. These trials would never have been granted ethical approval in the USA or France, the countries of the sponsors and collaborator. They exploited local regulatory weaknesses and economic and social inequities. They were pushed, approved and accepted by the sponsors (NIH and BMGF in the USA) and collaborator (IARC in France) to be conducted in India on poor and vulnerable women.

In these three trials, rights of the women participants in the no-screening control arm were violated: the universal and fundamental right to life and the right of access to the highest available standard of care. It was known that as poor women, they were already at increased risk of developing cervical cancer, and denying them known effective and potentially lifesaving screening put them at a predictable risk of developing invasive cervical cancer and dying from it. The denial of screening delayed not only the detection of the disease, but also access to appropriate and timely treatment that could have saved their lives. The standard of care was wrongly construed by the researchers as meaning the universal availability of tests under a programme of the government in India, rather than the universal standard of care used for testing of the disease, which was available in India.

For the purpose of public health policy, there was no need for a natural history control arm with no screening. The researchers should have provided an active control arm using one of the known methods of testing for cervical cancer, as they would have had to if the trials had been conducted in the USA or France.

In addition, the trials ignored the importance of informed consent. Women in the trials were not given adequate information. This violated their right to life, vitiated their consent and rendered the trial highly unethical. A trial without the participants' voluntary and informed consent would not have been permitted in an HIC.

What made these unethical trials possible? What were the conditions that enabled ethics dumping in the VIA trials? One needs to understand why host countries seek international support for research, why sponsors fund this research, and whether these reasons are justifiable. These reasons may include: a shortage of locally available funds for research; the interest of organizations in HICs in conducting research in LMICs as part of their international health agendas; and the relationships between local institutions and international organizations, as well as researchers' own links with these organizations as part of their individual scientific careers. All these create a web of relations that lies at the heart of the resulting double standard.

Research ethics must also contend with the view (Prasad et al. 2016) that locally relevant research justifies lower ethical standards. The researchers in these studies have argued that these studies are important because cervical cancer affects and kills poor women in LMICs more than it does women in HICs, and this calls for a test that is inexpensive, implementable and effective. They have also asserted that double standards do not cause active harm, as there is no functional screening system in the host country.

Finally, the women participants in both experimental and control arms of these trials are poor, voiceless and invisible. They may view participation in such trials as giving them access to some care. When faced with a powerful medical establishment, they are reluctant to make their grievances public. For instance, the hospital conducting the Osmanabad trial is the only such service in the area. In such a situation, violations of research ethics are less likely to come into the public eye.

Ethical Implications of Research in Communities Without Universal Access to Health Care

Researchers in the VIA trials did not provide the standard of care to participants in the control arm, arguing that India did not have an effective universal screening programme and its standard of care for cervical cancer prevention was therefore "no care".

Most LMICs, barring a few honourable exceptions, do not have *universal* access to health care. Even when the government is supposed to provide free access to health care, individuals are frequently forced to seek care in the private sector and pay for it. The care that people receive is therefore determined not by a universal standard but by what they can afford, or what the government provides, and many people do not get any care whatsoever. This situation has permitted researchers to interpret the standard of care, and their own responsibilities as physician-researchers, in a way that is not in the best interests of research participants.

The standard of care *cannot* depend on, or be defined according to, whether or not it is universally accessible. In the VIA trials, the Pap smear is the universal standard of care because it is universally considered to be an effective screening test for cervical cancer. Any woman who goes to a private or public hospital should expect to be offered it. It is part of the Indian government's cancer prevention programme.

Whether or not the community involved in research has universal access to the standard of care, through the government or private or social insurance, researchers and sponsors must be held responsible for providing this standard preventive, diagnostic and curative care free of cost to participants in the control arm of a trial. There is therefore a need to have an explicit provision in ethics guidelines and in the law emphasizing researchers' ethical obligation to provide standard care to participants in the control arm, as they are under their direct care during the course of research.

Regulatory Weaknesses

Guidance for ethics review of non-drug trials is included in the ethical guidelines of the Indian Council of Medical Research (ICMR) for biomedical research on human participants (ICMR 2006). The ICMR guidelines acknowledge that the denial of available treatment to a control group is unethical. They also state that "proper justification should be provided for using the placebo" and that "[i]n keeping with the Declaration of Helsinki *as far as possible standard therapy should be used in the control arm*" (ICMR 2006) (emphasis added).

However, since the trials were non-drug related, prior permission from the Drugs Controller General of India was not required. Thus the VIA trials did not have any legal oversight. The regulatory roles were played by institutional committees – the institutional ethics committees, scientific review committees and data safety monitoring committees. In this case, their authority was limited to within the institution and they were not accountable to a regulatory authority.

US regulatory bodies claimed inability to investigate and act on complaints of unethical research in the Osmanabad and Dindigul trials as these were funded by a private foundation. Hence, these trials were not accountable to the US regulator as they were not government-funded. Private foundations in HICs fund a substantial amount of collaborative research in low-income countries, and their lack of accountability to any authority is a matter of concern.

In the case of the Mumbai trial, the US regulatory body applied double standards. The use of a retrospective waiver of written informed consent, or permission to obtain consent after the intervention, goes against the very principle of prior informed consent in research, and would not have been allowed in the US. Likewise, the US OHRP did not conclude that the no-screening arm in the Mumbai trial was ethical, although it would not have been possible in the US. The trial even continued when the relevant local hospital ethics committee in Mumbai stated that the use of a no-screening arm was unethical.

Information about the actual trials, apart from the published papers, was not readily available. This prolonged the harm done to the participants, as it delayed the response to claims about the unethical and illegal nature of the trials.

Recommendations

The following steps are necessary to prevent ethics dumping between HICs and LMICs.

- *Ensure the regulation of collaborative research*. Studies involving international collaboration should only be allowed in LMICs if mechanisms are in place which ensure that the rights of participants will be respected at all times, and that sponsors, researchers, ethics committees or institutions, whether governmental or private, operating both inside and from outside the host country, are held accountable for their activities in the LMIC.
- *Ensure a framework for transparency*. Mechanisms must be put in place to ensure that trials are conducted in an open and transparent manner, and information about ongoing trials must be available and open to expert scrutiny, so as to prevent harm at any stage of the trial. The anonymized data, findings and conclusions of the researchers should be open to scrutiny, so that the findings and decisions on whether they should be used in public health policy can be properly evaluated.
- *Provide compensation for research-related injury*. Mariner (1997) writes:

> Since most legitimate research is intended to benefit society as a whole, the subject assumes risk for society's sake (some would say making a gift to society). Therefore, society has a moral obligation to make the injured subject whole by compensating those who took the risks and suffered thereby. In addition, it may be argued that where society conducts, supports, or sponsors research, it voluntarily assumes an obligation to compensate those who are injured in its enterprise.

Sponsors and researchers must compensate participants who suffer from trial-related injuries, by offering diagnostics and treatment freely and by providing monetary compensation for loss, injury, harm, mental and physical suffering, and expenses incurred as a result of participating in the trial. The mechanism should be simple, so that it causes minimal problems to the participants. In the above trials, proper follow-up of the women in the control arms, testing them with the best known methods, and providing treatment and compensation would be a step in the right direction. Families of women who died due to the standard of care being withheld, thereby preventing them from accessing timely treatment, must also be compensated.

References

ASCO Post (2013) ASCO 2013: Cervical cancer screening using visual inspection with vinegar reduces mortality by 31% in large study in India. The ASCO Post, 6 June. http://www.ascopost.com/News/4202

CIOMS (2002) International ethical guidelines for biomedical research involving human subjects. Council for International Organisations in Medical Sciences, Geneva

Ferlay J, Soerjomataram I, Ervik M, Dikshit R, Eser S, Mathers C, Rebelo M, Parkin DM, Forman D, Bray F (2013) GLOBOCAN 2012: Estimated cancer incidence, mortality and prevalence worldwide in 2012 v1.0. IARC CancerBase No. 11. International Agency for Research on Cancer, Lyon. http://globocan.iarc.fr

Howell TR, Ballantine D (1998) Dumping: still a problem in international trade. In: Wessner CW (ed) International friction and cooperation in high-technology development and trade: papers and proceedings. National Academy Press, Washington CD, p 325–377

ICMR (2000) Ethical guidelines for biomedical research on human participants. Indian Council of Medical Research, New Delhi. http://web.archive.org/web/20010613180317/http://icmr.nic.in/ethical.pdf

ICMR (2006) Ethical guidelines for biomedical research on human participants. Indian Council of Medical Research, New Delhi. http://www.icmr.nic.in/ethical_guidelines.pdf

Krishnan S, Madsen E, Porterfield D, Varghese B (2013) Advancing cervical cancer prevention in India: insights from research and programs. Health, nutrition and population discussion paper. World Bank, Washington DC

Macklin R (2004) Double standards in medical research in developing countries. Cambridge University Press, Cambridge

Mariner W (1997) Compensation for research injuries. In: Mastroianni AC, Faden R, Federman D (eds) Women and health research: ethical and legal issues of including women in clinical studies, vol 2: workshop and commissioned papers. Institute of Medicine and National Academy Press, Washington DC, p 113–126

Mittra I, Mishra GA, Singh S, Aranke S, Notani P, Badwe R, Miller AB, Daniel EE, Gupta S, Uplap P, Thakur MH, Ramani S, Kerkar R, Ganesh B, Shastri SS (2010) A cluster randomized, controlled trial of breast and cervix cancer screening in Mumbai, India: methodology and interim results after three rounds of screening. International Journal of Cancer 126(4):976–984. doi:10.1002/ijc.24840

National Cancer Control Programme (2006) Guidelines for cervical cancer screening programme: recommendations of the expert group meeting. Government of India – World Health Organization Collaborative Programme (2004–2005). Department of Cytology & Gynaecological Pathology, Postgraduate Institute of Medical Education and Research, Chandigarh, India. http://screening.iarc.fr/doc/WHO_India_CCSP_guidelines_2005.pdf

Pramesh CS, Shastri S, Mittra I, Badwe R (2013) Ethics of "standard care" in randomised trials of screening for cervical cancer should not ignore scientific evidence and ground realities. Indian Journal of Medical Ethics 10(4):250–251

Prasad V, Kumar H, Mailankody S (2016) Ethics of clinical trials in low-resource settings: lessons from recent trials in cancer medicine. Journal of Global Oncology 2(1): 1–3

Sankaranarayanan R, Budukh AM, Rajkumar R (2001) Effective screening programmes for cervical cancer in low- and middle-income developing countries. Bull World Health Organ 79 (10):954–962

Sankaranarayanan R, Esmy PO, Rajkumar R, Muwonge R, Swaminathan R, Shanthakumari S, Fayette JM, Cherian J (2007) Effect of visual screening on cervical cancer incidence and mortality in Tamil Nadu, India: a cluster-randomised trial. Lancet 370(9585):398–406

Sankaranarayanan R, Nene BM, Dinshaw K, Rajkumar R, Shastri S, Wesley R, Basu P, Sharma R, Thara S, Budukh A, Parkin DM (2003) Early detection of cervical cancer with visual inspection methods: a summary of completed and on-going studies in India. Salud Pública de México 45 (suppl. 3):S399–S407

Sankaranarayanan R, Nene BM, Dinshaw KA, Mahe C, Jayant K, Shastri SS, Malvi SG, Chinoy R, Kelkar R, Budukh AM, Keskar V, Rajeshwarker R, Muwonge R, Kane S, Parkin DM, Chauhan MK, Desai S, Fontaniere B, Frappart L, Kothari A, Lucas E, Panse N (2005) A cluster randomized controlled trial of visual, cytology and human papillomavirus screening for cancer of the cervix in rural India. International Journal of Cancer 116(4):617–623

Sankaranarayanan R, Nene BM, Shastri SS, Jayant K, Muwonge R, Budukh AM, Hingmire S, Malvi SG, Thorat R, Kothari A, Chinoy R, Kelkar R, Kane S, Desai S, Keskar VR, Rajeshwarkar R, Panse N, Dinshaw KA (2009) HPV screening for cervical cancer in rural India. New England Journal of Medicine 360(14):1385–1394

Sankaranarayanan R, Nene BM, Shastri SS, Jayant K, Muwonge R, Malvi SG (2011) Reply to S D Rathod's Commentary on HPV screening for cervical cancer in rural India. Indian Journal of Medical Ethics 8(3):182–183

Sankaranarayanan R, Rajkumar R, Rajapandian T, Pulikattil OE, Mahe C, Bagyalakshmi K, Thara S, Frappart L, Lucas E, Muwonge R, Shanthakumari S, Jeevan D, Subbarao TM, Parkin DM, Cherian J (2004) Initial results from a randomized trial of cervical visual screening in rural south India. International Journal of Cancer 109:461–467

Shastri SS, Mittra I, Mishra GA, Gupta S, Dikshit R, Singh S, Badwe RA (2014) Effect of VIA screening by primary health workers: randomized controlled study in Mumbai, India. Journal of the National Cancer Institute 106(3):dju009. doi:10.1093/jnci/dju009

Suba EJ (2014) US-funded measurements of cervical cancer death rates in India: scientific and ethical concerns. Indian Journal of Medical Ethics 11(3):167–175

WMA (2000) WMA Declaration of Helsinki: Ethical principles for medical research involving human subjects. World Medical Association. http://www.who.int/bulletin/archives/79(4)373.pdf

WMA (2008) WMA Declaration of Helsinki: Ethical principles for medical research involving human subjects. World Medical Association. http://ethics.iit.edu/ecodes/node/4618

Author Biographies

Sandhya Srinivasan is a Mumbai-based freelance journalist and researcher. She was executive editor of the *Indian Journal of Medical Ethics* from 1998 to 2011 and is now consulting editor. She is also consulting editor for scroll.in, a digital daily of political and cultural news in India, and a trustee of the Centre for Communication and Development Studies, a non-profit organization advocating for social justice and sustainable development in India.

Veena Johari is a Mumbai-based lawyer working on human rights and public health. She was with the Lawyers' Collective HIV/AIDS Unit before setting up Courtyard Attorneys, which focuses on access to medicines and opposition to patents on pharmaceutical drugs. She is the legal expert in the research ethics committee of the King Edward Memorial Hospital in Mumbai, and is on the editorial board of the *Indian Journal of Medical Ethics*.

Amar Jesani is an independent consultant, researcher and teacher in bioethics and public health. He is co-founder of the Forum for Medical Ethics and current editor of the *Indian Journal of Medical Ethics*. He is visiting professor at the Centre for Ethics at Yenepoya University, Mangalore, and associate faculty at the Centre of Biomedical Ethics and Culture, Karachi. He is a member of the Institutional Review Board at the Tata Institute of Social Sciences, Mumbai, the International Research Ethics Committee of Médecins Sans Frontières, and the Central Ethics Committee of the Indian Council of Medical Research.

Chapter 6
Ebola Vaccine Trials

Godfrey B. Tangwa, Katharine Browne and Doris Schroeder

Abstract The Ebola epidemic that broke out in West Africa towards the end of 2013 had been brought under reasonable control by 2015. The epidemic had severely affected three countries. This case study is about a phase I/II clinical trial (testing for safety and immunogenicity) of a candidate Ebola virus vaccine in 2015 in a sub-Saharan African country which had not registered any cases of the Ebola virus disease. The study was designed as a randomized double-blinded trial. It was sponsored and funded by one of the biggest Northern multinational pharmaceutical companies. The protocol received ethics clearance from the relevant national ethics committee. The study was coordinated and managed at the local branch of a big Northern diagnostic laboratory and a laboratory of a local regional hospital. The overall study was a multi-country, multi-site trial aimed at recruiting a total of 3,000 research participants across four or five sub-Saharan African countries. For this country, the recruitment sites were two big cities, each aiming to recruit 200 participants: adults at the first site and children at the second. The target sample size was almost achieved at the first site but, before the study commenced at the second site, some members of (the public) raised the alarm that the government was carelessly risking the health, safety and lives of citizens in the cause of an unproven vaccine that could precipitate a public health disaster. The trial was immediately suspended. A commentary on this case, and on the importance of trust, is provided by Katharine Browne and Doris Schroeder at the end of this chapter. It highlights differences between this case and a phase I Ebola vaccine trial in Canada in 2014.

G. B. Tangwa (✉)
Cameroon Bioethics Initiative (CAMBIN), University of Yaounde 1,
P.O. Box 13597, Yaounde, Cameroon
e-mail: gbtangwa@gmail.com

K. Browne
Department of Philosophy, Langara College, 100 West 49th Avenue,
Vancouver B.C., CanadaV5Y 2Z6

D. Schroeder
Centre for Professional Ethics, University of Central Lancashire,
Brook 424, Preston PR1 2HE, United Kingdom

D. Schroeder et al. (eds.), *Ethics Dumping*, SpringerBriefs in Research
and Innovation Governance, https://doi.org/10.1007/978-3-319-64731-9_6

Keywords Ebola · Candidate Vaccine Trial · Sub-Saharan Africa
Ethics committee · Trust

Area of Risk of Exploitation

Phase I clinical trials are trials in which the safety of a new treatment is tested in a small group of individuals (often healthy volunteers) to evaluate safety and side effects and to determine dosage. The chances of therapeutic outcomes for the research participants are almost always zero. In this context, the risk of exploitation of low- and middle-income country (LMIC) participants is particularly high, as, due to low education levels, they are more likely to assume that they will benefit personally. For this reason, phase I clinical studies have previously been carried out only in high-income countries. However, they are now increasingly also carried out in LMICs, especially in accordance with community engagement procedures and where the expected outcomes of the study mostly or exclusively benefit LMICs. The same applies, with limitations, to phase II clinical trials, whose main purpose is to assess efficacy. In phase I/II clinical trials in LMICs, it is therefore particularly important to protect research participants.

Specific Case

This case study is about a phase I/II clinical trial (testing for safety and immuno-genicity) of a candidate Ebola virus vaccine in a sub-Saharan African country in 2015.

In early 2015, a team of experts from the country's Ministry of Public Health evaluated the availability of facilities for Ebola vaccine trials in certain medical centres and laboratories. The team included members of the country's National Ethics Committee (NEC). Equipment inventoried during this visit, in at least one of the centres, was marked as a previous donation from the owner of the candidate vaccine. After these visits, two urban medical centres were retained for the Ebola vaccine study.

The visits occurred after the Ebola epidemic that had broken out in West Africa towards the end of 2013 and that had been brought under reasonable control by 2015. The epidemic had severely affected three countries. However, the country of the candidate vaccine trial had not registered any cases of the Ebola virus disease.

The study was designed as a randomized double-blinded trial in which half of the research participants would receive the candidate vaccine and the other half a placebo. The study was sponsored and funded by one of the biggest Northern multinational pharmaceutical companies, globally well known and highly respected. The first medical centre was to recruit 200 adult research participants and the second 200 children. The protocol of the study, at least for the first recruitment site, received ethics approbation from the NEC. Recruitment was nearing completion at the first site when, following complaints from some members of the public, the study was suspended.

The suspension order was apparently made by word of mouth. The Minister of Public Health who had initially announced the commencement of the study over the radio did not announce its suspension through any public media. However, he did write to the principal investigator (PI) at the local branch of the Northern diagnostic laboratory to explain that the study had been suspended due to public protests and that the trial involving children would now be withdrawn too. The general public, as well as the research participants and their families and communities, knew little about the study, let alone why it had been suspended, and therefore permitted themselves the most fanciful speculation about it.[1]

Case Analysis

This case bristles with ethical problems and issues that go beyond any simple identification of instances of North-South "ethics dumping". It involves subterfuges to circumvent standard regulatory procedures and discretion bordering on secrecy – approaches that would be inconceivable in high-income countries, or anywhere else where there is sufficient awareness of the stakes of biomedical research and the ethics of clinical research, particularly that involving human research participants.

The risk of exploitation in this case is not limited to a single rubric such as "no benefit sharing" or "inadequate informed consent process", but relates rather to the exploitation of the general weaknesses and inadequacies of an entire system, particularly its lack of a credible and adequate research governance and regulatory framework. This suggests double standards that would also be inconceivable in a high-income country.

International regulatory texts are simple and clear on the procedural rules for the ethically acceptable conduct of medical research, particularly clinical trials, with human beings as research participants. The *Declaration of Helsinki*, for instance, states:

> The research protocol must be submitted for consideration, comment, guidance and approval to the concerned research ethics committee before the study begins. This committee must *be transparent in its functioning*, must be *independent of the researcher*, the sponsor and any other undue influence and must be *duly qualified* (WMA 2013: art. 23) (emphasis added).

Regarding "post-trial provisions", the declaration states:

> In advance of a clinical trial, sponsors, researchers and host country governments should make provisions for post-trial access for all participants who still need an intervention

---

[1]For popular concerns raised in such contexts, see Geissler and Pool (2006).

identified as beneficial in the trial. This information must also be disclosed to participants during the informed consent process (WMA 2013: art. 34).

These minimal conditions were evidently not fulfilled for this trial. The members of the ethics committee that approved the study were all appointees by decree of the Minister of Public Health and functioned within a civil service system where obedience to hierarchical superiors was regarded as a duty. It is the view of this author that by assuming sponsorship of the clinical trial in this case, the Minister virtually made it the duty of all within the Ministry of Public Health to help facilitate its accomplishment. This would explain why some members of the ethics committee were involved in prior site preparations for the trial, which was inappropriate for an ethics committee as it compromised ethical oversight of the study and put the independence and transparency of the ethics committee in serious doubt.

There is no doubt that all the members of the approving ethics committee (which in fact acts as the national ethics committee) are highly qualified in their fields, but this does not automatically make them experts in ethics review. The expertise represented in the committee is roughly as follows: a haematologist, a parasitologist/epidemiologist, a pneumological epidemiologist, a sociologist, a demographer, an x-ray oncologist, a pathologist, a jurist, a parasitologist, a surgeon, a microbiologist/pharmacist, a dental surgeon, an expert in the science of education, a paediatrician, a civil society member, a traditional practitioner, an expert on Islamic religion and two community members. This is a highly impressive committee for science review, perhaps, but not necessarily for ethics review, if no research ethics training has been provided. And even if such training has been provided, which usually happens by way of workshops or symposia, every research ethics committee still needs an ethics expert, meaning someone whose main business and concern as a member of the committee is ethics aspects of and ethics issues in the protocol.

The study was suspended before the recruitment of children had begun at the second site. The inclusion of children in a clinical study designed for testing safety and immunogenicity is a big ethical issue for which, at best, no justification was available for this study. All over Africa, women and children, because of their vulnerability, high rate of morbidity, easy availability, naivety and trustfulness, bear a heavy burden of clinical research. A competent ethics committee would have checked the burden of research participation against the benefits for research participants and their immediate communities, especially where children were involved.

The whole study involved structures and procedures that, on the surface, appeared to conform to ethics demands but, in reality, violated the principles of research participant protection that are paramount in research ethics. The following section analyses the case further, with specific reference to the informed consent documentation.

The Informed Consent Process

The potential research participant information for this study contained inadequacies and issues that any qualified ethics committee should have noticed and raised with the investigators for redress or amelioration before subject recruitment commenced.

Regarding the informed consent process, the *Declaration of Helsinki* states:

> [E]ach potential subject must be <u>adequately informed</u> of the aims, methods, sources of funding, any possible conflicts of interest, institutional affiliations of the researcher, the anticipated benefits and potential risks of the study and the discomfort it may entail, post-study provisions and any other relevant aspects of the study (WMA 2013: art. 26) (emphasis added).

An analysis of the information sheets given to the potential participants in this clinical trial shows serious omissions and inappropriate or misleading language for the context (see also below). Participants were not taken through any informed consent process other than being approached individually by the study physicians or their agents at the chosen site of the study, given the information sheets to take home and asked to come back the next day to sign the informed consent form, followed by procedural instructions, and payment to them of a sum of approximately US$20. Furthermore, there was no process of community engagement beyond the media announcement by the Minister of Public Health emphasizing that the study had been approved by the World Health Organization (WHO) and was simultaneously taking place in many countries. In retrospect, the Minister's announcement could be judged an inducement which played down the potential stakes and risks of the study by referring to the approval of the WHO and the fact that other countries had accepted the trial.

Some of the issues and questions addressed in the five-page prospective participants' information sheet are quoted below, in italics. Each excerpt is followed by my attempt to review and critique it in the way a competent and vigilant ethics committee might have done.

Why is this clinical study being done?

This study is done to test a vaccine against Ebola to make sure that it is safe and that it brings about a protective response. You can get Ebola by being in direct contact with the blood or other body fluids of a person who is already sick with Ebola. People infected with Ebola can have many different symptoms, like fever, severe headache, muscle pain, weakness, feeling tired, diarrhoea, vomiting, stomach pain and unexplained bleeding or bruising. Ebola disease can be very severe and is a life-threatening disease.

Only the first sentence of the above response addresses the question asked. But it is misleading and confusing to state that the study is being done "to test a vaccine against Ebola" whereas the answer to a subsequent question below states that "there are no vaccines or treatment against the Ebola virus". The candidate vaccine ought to be called and described accurately for what it is. The rest of the response is not relevant to the question: it answers another question that has not been asked, namely: what is Ebola and how does one get it?

Who can take part in this study?

You can take part in this study if you are at least 18 years old, healthy, not taking part in another study, have not been in a country affected by the Ebola epidemic (Sierra Leone, Liberia and Guinea) and have not been in contact with someone who has Ebola in the last 3 weeks.

In the country of this study, the age of majority and consent is 21. Since clinical studies need to conform to local laws and regulations, the age of participation here should be 21, not 18; or else it should be explained that those below 21 would require the proxy consent of their parent or legal guardian in addition to their own assent.

Considering the issue of fairness in the recruitment of study participants, it may be questionable to make the mere fact of having been to an Ebola country and even of having been in contact with an Ebola patient – without, however, contracting the disease – an exclusion criterion for the study.

Which vaccine will you get?

You will get the Ebola vaccine, either at the start of the study, or after 6 months into the study. At the start of the study, half of the people in the study (about 1,500 people) will get the Ebola vaccine and the rest of the people will get a placebo (dummy vaccine that looks like a real vaccine but does not have active components in it). Neither you nor the study doctor can choose or will know which vaccine you receive. This will be randomly decided by a computer (like the flip of a coin). We will only tell you and the study doctor which vaccine you received after 6 months into the study, or if there is an emergency.

This response repeatedly refers to an "Ebola vaccine", even though it does not yet exist. It also fails to explain in simple ordinary language such terms as "placebo", "dummy vaccine" and "active components", and to illustrate what may count as an "emergency". Informed consent processes must avoid jargon, especially in LMIC settings.

What does this study involve?

The first study visit, called a Screening visit, is to check if you can take part in the study. The study doctor will ask you some questions, do a physical examination and take some blood to test blood factors. If you are a woman who can get pregnant, the study doctor will also ask for a pregnancy test. If the Screening shows that you can take part in the study, you will be in the study for about 1 year. Half of the people to join the study will have extra study procedures done. ... if you are part of the group of people that does NOT need extra procedures you will: visit the vaccination centre 4 times after the Screening on Day 0, month 1, month 3 (phone call/home visits), month 6, month 9 (phone call/home visit) and month 12. At Day 0 visit, you will be vaccinated. This is an injection in the muscle of your upper arm. After vaccination, you need to stay at the vaccination centre for at least 30 minutes for observation. If you receive the dummy vaccine at the first visit, you will be vaccinated with the Ebola vaccine 6 months later. ... During the entire study, we will check if you have any serious medical conditions. You will not have blood taken during the rest of the study.

First, here too the use of technical medical terms is problematic ("screening visit", "blood factors", "extra study procedures", "muscle of your upper arm", "serious medical conditions"). Second, this may make a good entry in the notebook of the investigator but will not necessarily be meaningful to a prospective subject without prior verbal explanation. What, for instance, is a barely literate person to make of "Day 0, month 1, month 3 (phone call/home visits)"? Would s/he not be wondering how a day could be zero and if s/he would be required to telephone someone or visit them at home?

What about pregnancy?

We do not know yet if the Ebola vaccine may have an effect on an unborn baby. That is why you should not take part in this study if you are pregnant or trying to get pregnant. ... You will need to use birth control during the first 7 months you take part in this study. Tell the study doctor if you are pregnant during the first 7 months of the study. The doctor will follow you up until the delivery of the baby.

This explanation about pregnancy is not free of ambiguity. It is quite clear that I should not take part in the study if I am pregnant or want to get pregnant. It is also clear that, if I want to have sex during the seven months of the study, I should use contraception. But telling me that I should inform the study doctor if I get pregnant during the first seven months of the study so that s/he can follow me up until the delivery of my baby is rather confusing. A study participant might say: "Getting pregnant and being followed up until delivery is what I want most. So why does the informed consent documentation say that I should not take part in the study if I am pregnant or want to get pregnant?" This paragraph crucially fails to explain that contraception can sometimes fail because no method of contraception or birth control is 100% effective except abstinence.

What benefits can you expect?

You may not benefit from the Ebola vaccine because we do not know yet if the vaccine will be able to protect people against Ebola virus.

This is at best an incomplete response to the question. Of course you do not know yet if the vaccine will be able to protect people against the Ebola virus; that is the whole purpose of the trial test. But what happens if/when it does prove to be able to protect people against the Ebola virus disease? How is article 34 on "Post-Trial Provisions" of the *Declaration of Helsinki* going to be respected?

What side effects or risks can you expect?

There is a very small risk that you could have an allergic reaction after vaccination. ... That is why it is important that you stay at the vaccination centre for at least 30 minutes after vaccination, where all medical tools are available to treat an allergic reaction.

To avoid misunderstanding among research participants with low literacy and education levels, it would be better to rephrase this response in terms of the possibility, not how small the risk – "It is possible that you could have an allergic reaction" – followed by an explanation of what an allergic reaction is.

Are there other vaccines or treatments?

So far, there are no vaccines or treatment against the Ebola virus.

For clarity, the question should be "Are there other vaccines or treatments against Ebola?" Also note the confusion if one compares this statement with earlier mentions of a vaccine, for instance in "You will get the Ebola vaccine ..." (see above).

What happens if you leave the study?

If you leave the study we will keep and use the information and samples we collected before you left the study. We will ask you to return to the vaccination centre one more time for a safety follow-up.

The obvious follow-up questions not addressed here are: Why would you keep and use the information and samples you collected even when I have decided to leave the study? And why should I return to the vaccination centre again after I have decided to leave? What safety follow-up are you talking about?

Who will be looking at the information from this study?

Your information will be protected in accordance with the most stringent applicable law.

When you sign/thumb print this consent form you agree that your information can be viewed and used by site staff, [the pharmaceutical company], agencies and independent ethics committee. ... [the pharmaceutical company] may publish the results but your name will not appear in any publication. If you withdraw consent to use your personal information you will no longer be able to continue in the study.

The questions that need addressing here are: what is "the most stringent applicable law" that will protect my information, and why should my name not appear in any publication of the results in spite of my contribution to it? (See also the supplementary report after this case study, which describes the pride with which Canadian research participants in a phase I Ebola vaccine trial made their names public.)

What happens if you get injured while taking part in this study?

If you are harmed by the vaccination in the study or by any of the study procedures, you will be compensated. Your study doctor can give you information about how to obtain compensation in case of injury. You will not be paid for taking part in this study but you will be paid reasonable travel fees to attend to study visits at the vaccination centre.

This question needs a fuller and clearer answer. Compensation for study-related injury should not vaguely be referred to the study doctor; it should be explained clearly. "You will not be paid... but you will be paid..." is not a good formulation in informed consent information and needs to be rephrased less ambiguously or even misleadingly.

The informed consent form (certificate) states:

The study has been explained to me. I have read the information or have had the information read to me. I have been given enough time to make a decision. I have had the chance to ask questions and I am happy with the answers that I have been given. I have been told that I can change my mind at any time and stop taking part in the study without giving any reason. By signing/thumb printing this form I agree:

1. To take part in the study
2. That my information is used as described in this form
3. That my blood samples are used as described in this form

Tick as appropriate (this decision will not affect your ability to take part in the study):

YES. My samples may also be used for future research (at the time of the study or after the study is finished) not described in this form with prior approval of the Ethics Committee
NO. Do not use my samples for future research (at the time of the study or after the study is finished) not described in this form.

The tickable options above are about something as important as the use of samples for unknown future research. This ought to be discussed and justified in information designed for the prospective participant. As formulated here, the section in parentheses is not clear and is liable to be quite confusing: "at the time of the study or after the study is finished" should perhaps be changed simply to "after this study is finished".

Talking to two of the potential research participants suggested that the main motivations for participation were the incentives – the health care benefits and the money paid. Of course, there are limits to the conclusions one can draw from talking to only two people, but the literature shows that financial incentives and access to health care are a major driver for enrolment in studies in LMICs (Mfutso-Bengo et al. 2008; Mduluza et al. 2013). For this reason the case raises concerns about undue inducements.

Conclusion

The regulation of human subject research and particularly of clinical research is quite advanced around the globe, to the extent that we can talk about a regulatory infrastructure, whose presence or absence in any given context should indicate a priori whether or not research involving human subjects can ethically be conducted within that context. Such regulatory infrastructure would include, in a non-authoritarian and genuinely democratic context, a legal framework that respects fundamental human rights, especially freedom of inquiry and expression, overseen by well-constituted, qualified and genuinely independent ethics review committees.

The absence of such infrastructure or doubt about its genuineness, in spite of appearances, delimits a no-go area for ethical research. The verifiable existence of such an infrastructure should be a precondition for human subject research, especially in resource-destitute settings and particularly in the first two phases of investigation. Transparency must be part and parcel of any procedures where publicly available regulations need to be followed. Systemic faults tend to render compliance with good procedural rules and practices not only difficult, but impossible. It is not just difficult, but impossible, to carry water in a straw basket for any distance.

Supplement to the Ebola Vaccine Trial Case – The Importance of Trust

This is an excerpt from a Canadian newspaper article:

> Hundreds of Nova Scotians are volunteering to be injected with an experimental vaccine that might cause aches and fever – but could protect against the Ebola virus.
>
> Within minutes of the Nov. 14 announcement that Halifax's IWK Health Centre was chosen to hold the clinical trial to test Canada's Ebola vaccine, the phones started ringing and e-mails began arriving from people who wanted to participate. A week later, the trial team has heard from about 300 people – it only needs 40 healthy individuals, between 18 and 65, for this first-phase trial (Taber 2014).

The phase I trial for the Canadian-developed Ebola vaccine (VSVΔG-ZEBOV) was conducted at the Izaak Walton Killam Health Centre in Halifax, Nova Scotia. The trial involved 11 clinic visits over six months, each requiring a blood draw. Participants received CAD 1,125 for their participation in the entire trial.

One of the authors of this supplement, Katharine Browne, is involved in a study examining the factors that motivate healthy volunteers to participate in phase I vaccine trials. The study involves a survey of the motivations of healthy volunteers for the Canadian phase I Ebola vaccine trial, as well as a phase I trial for a PAL adjuvant.[2] The central hypotheses of the study are that:

1. The financial incentive will be the dominant motivation that participants identify.
2. Other motivations will include a desire to contribute to the development of a vaccine, and a desire to help others.
3. The high-profile nature of the Ebola vaccine trial will play a factor in participant motivations.

Surprisingly, and contrary to the first hypothesis, preliminary findings from the study reveal that financial incentives are neither the sole nor the main determinants in motivating individuals to participate in vaccine trials. The findings do, however, confirm the second hypothesis: that participant motivations include desires to help develop a new vaccine and to help others. One research participant explained to the media that participating in the trial had been a life-changing experience for her (CTVNews.ca Staff 2014). When asked about the financial incentive provided for participation, she said that she would put it towards her university studies. She also noted that another research participant had donated the money he received for participating to a children's charity (CTVNews.ca Staff 2014). Concerning the third hypothesis, the findings are unable to confirm or deny that the high-profile nature of the Ebola vaccine trial contributed to trial participation.

---

[2]An adjuvant is an immune booster that can be added to a vaccine. PAL is the name of a particular adjuvant.

The study findings, along with the anecdotes from trial participants, support a general trend away from the selfish actor model that underlies classical economic theory and that informs policies and practice, including payment for research participants.

The Canadian experience of the phase I Ebola vaccine trial provides a remarkable contrast to the almost identical study at the African site. One possible explanation for the over-recruitment at the Canadian site compared with the public outcry at the African site could be the extent to which the two trials differ in the levels of trust between research participants and researchers. The suggestion here is that there is a lack of trust in North-South collaborations and that this dramatically affects the recruitment of research participants. Further research is required to confirm or deny this hypothesis. To enhance trust in such collaborations, we re-emphasize Tangwa's two main conjectures:

1. The verifiable existence of an infrastructure that respects fundamental human rights should be a precondition for medical research involving human participants, especially in LMICs and particularly in the first two phases of investigation.
2. Transparency is essential when conducting trials in LMICs.

In addition, Tangwa's analysis of the informed consent documentation reveals a notable ignorance of local requirements (e.g. researchers seeming unaware of the local age of consent). This is likely to contribute to distrust in North-South collaborations.

One could venture that the non-existence of a reliable governance structure and non-transparency, combined with insensitivity to local requirements, have a major impact on trust.

References

CTVNews.ca Staff (2014) Canadian who received Ebola vaccine says experience has changed her. CTV News, 29 December. http://www.ctvnews.ca/canadian-who-received-ebola-vaccine-says-experience-has-changed-her-1.2164898

Geissler PW, Pool R (2006) Popular concerns about medical research projects in sub-Saharan Africa: a critical voice in debates about medical research ethics. Tropical Medicine and International Health (editorial) II(7):975–982

Mduluza T, Midzi N, Duruza D, Ndebele P (2013) Study participants incentives, compensation and reimbursement in resource-constrained settings. BMC Medical Ethics 14(suppl 1): S4

Mfutso-Bengo J, Ndebele P, Jumbe V, Mkunthi M, Masiye F, Molyneux S, Molyneux M (2008). Why do individuals agree to enrol in clinical trials? A qualitative study of health research participation in Blantyre, Malawi. Malawi Medical Journal 20(2):37–41

Taber J (2014) Hundreds of Nova Scotians volunteer for Ebola vaccine trial. The Globe and Mail, 21 November. http://www.theglobeandmail.com/life/health-and-fitness/health/hundreds-of-nova-scotians-volunteer-for-ebola-vaccine-trial/article21712654/

WMA (2013) WMA Declaration of Helsinki: Ethical principles for medical research involving human subjects. World Medical Association. https://www.wma.net/policies-post/wma-declaration-of-helsinki-ethical-principles-for-medical-research-involving-human-subjects/

Author Biographies

Godfrey B. Tangwa is professor of philosophy and former head of the Department of Philosophy at the University of Yaounde. He is a member of the International Association of Bioethics, and was vice president from 1999 to 2001. He is a fellow of the Cameroon Academy of Sciences and the African Academy of Sciences, founder and chairperson of the Cameroon Bioethics Initiative, and advisory board member and chairperson of the Cultural, Anthropological, Social and Economic working group of the Global Emerging Pathogens Treatment Consortium.

Katharine Browne is a faculty member in the Department of Philosophy at Langara College, Vancouver. She is a past Research Ethics Board member at the Izaak Walton Killam Health Centre and current affiliated member of the Canadian Center for Vaccinology.

Doris Schroeder is director of the Centre for Professional Ethics at the University of Central Lancashire, and the School of Law, UCLan Cyprus, and adjunct professor at the Centre for Applied Philosophy and Public Ethics, Charles Sturt University, Canberra. She is coordinator of the TRUST project and has previously guided large international consortia on the topics of benefit sharing and responsible research and innovation.

Chapter 7
Hepatitis B Study with Gender Inequities

Olga Kubar

Abstract This case study is about a study entitled "Comparable randomized double-blind investigation of safety and immunogenicity of vaccine against Hepatitis B in healthy adult subjects" proposed in Russia with an international sponsor. There were indications of elements of exploitation, which consisted of inadequacies in the study's design compared with its announced purpose, and the indirect inclusion of women research subjects in the clinical trial without their informed consent. On the basis of noncompliance with the applicable regulatory and ethical requirements the study was not approved by the local ethics committee (LEC).

Keywords Clinical trial · Hepatitis B · Russia · Women · Exploitation Unethical · Ethics committee

Area of Risk of Exploitation

Healthy volunteers in clinical trials contribute to medical progress without any benefits to themselves. In addition, this case is of interest with regard to gender inequities in research.

Case Description

This case study is based on an evaluation undertaken by the local ethics committee (LEC) of the research institute in Russia at the end of 2014. All documentation required for a complete ethical review of the proposed study was submitted in accordance with the national law (Russian Federation 2005), the LEC's standard operating procedure and international rules of good clinical practice. The proposed

O. Kubar (✉)
Saint Petersburg Pasteur Institute, St. Petersburg 197101, Russia
e-mail: okubar@list.ru

© The Author(s) 2018
D. Schroeder et al. (eds.), *Ethics Dumping*, SpringerBriefs in Research and Innovation Governance, https://doi.org/10.1007/978-3-319-64731-9_7

61

clinical trial was entitled: "Comparable randomized double-blind investigation of safety and immunogenicity of vaccine against Hepatitis B in healthy adult subjects".

The main purpose of the proposed clinical trial was to study the safety and immunogenicity of a vaccine against hepatitis B in comparison with a vaccine already marketed in Russia, with a view to its future registration in the country.

The study design envisaged two groups of participants made up of both men and women. The first group would be vaccinated by an investigational product (a vaccine proposed by an external sponsor), and the second (control) group would be given a well-known vaccine registered in the country. According to the protocol, the female sexual partners of male participants would be indirectly involved. For this group, the study assigned special requirements.

The requirements for these women, who were not legally and directly involved in the clinical trial, included a prohibition on and prevention of pregnancy, through the use of contraception, during the entire eight months the study lasted and for one month afterwards, even if the actual participant – their sexual partner – withdrew from the study.

Detailed information would be collected about any pregnancy and its outcome, and any adverse events (or serious adverse events) would be included in the database as part of the monitoring process.

The investigational product had been well investigated in a series of earlier clinical trials (as is clear from the protocol, investigation brochure and references), and already approved in the country of the sponsor and many other high-income countries. It was available on the open market for adults and children above ten years old. For this reason, the appropriate design of the proposed clinical trial in Russia would have been for a phase III study. However, the protocol design was equivalent to a phase I or II study.

Seventeen visits of the volunteer participants to the investigator centre were planned during the eight months of the clinical trial and for one month after its completion. Visit procedures involved a detailed physical examination and the collection of blood and urine samples for a wide spectrum of tests. The participants would come to the centre in the morning and spend a few hours there for observation. In addition, they would have to buy and use the requested products for contraception.

As a rule, healthy volunteers participating in clinical trials cannot expect any benefits. In this case, an external (i.e. non-Russian) sponsor declared that benefits were planned, because the participants (volunteers) would be vaccinated against hepatitis B, and would therefore be protected from this infection in the future.

Analysis

The case study shows ethical inadequacy at several levels.

The suggestion that the study would be beneficial to participants is controversial. Vaccination against hepatitis B is included in the national immunization calendar of the Russian Federation. This is done with domestically and internationally produced vaccines that are registered and have been granted permission for use by approved order (Russian Federation 2014). Vaccination against hepatitis B is freely available to everybody, and obligatory for high-risk groups (newborns whose mothers are HbsAg carriers, or hepatitis B patients in the third trimester of pregnancy). Therefore there were no benefits for participants taking part in the clinical trial.

The autonomy of the women who were indirectly involved in the study was not respected. There was no information or confirmation in any part of the protocol to the effect that these women (indirect participants) should be appropriately informed about the procedures, or that their informed consent should be obtained.

In addition, their indirect involvement in the clinical trial was not covered by insurance, even in the case of pregnancy with a serious adverse event (a congenital anomaly or birth defect), because they were not included in the framework of financial contracts and insurance coverage for study participants. No other guarantee (medical, financial etc.) for these women was described in the protocol or any other study documents. This violates Russia's compulsory regulations on good clinical practice:

> In research which does not connect with treatment (without any benefits for potential participants from a medical point of view) only subjects who personally give, write and date the informed consent can be involved (Russian Federation 2005: item 4.8.13).

The situation for women indirectly involved in the clinical trial without consent would also contradict the universal ethical principles of the Declaration of Helsinki, October, 2013 regarding vulnerable groups and populations:

> Article 22: "The protocol should include information … regarding provisions for treating and/or compensating subjects who are harmed as a consequence of participation in the research study"

> Article 25: "Participation by individuals capable of giving informed consent as subjects in medical research must be voluntary … no individual capable of giving informed consent may be enrolled in a research study unless he or she freely agrees"

> Article 26: "The potential subject must be informed of the right to refuse to participate in the study or to withdraw consent to participate at any time without reprisal" (WMA 2013).

The requirement to carry out a pregnancy test and prevent pregnancy throughout the study also violated the women's reproductive rights and represented a direct intervention in the family's planning.

The study documentation required considerable attention to be devoted to the registering and following up of information concerning cases of pregnancy or outcomes in these women, without their informed consent. This meant that their personal information could be used without their agreement. It also contradicted the general norms guaranteeing the protection of personal data set by the Russian Federation's Federal Law on Information, Information Technologies and the Protection of Information (Russian Federation 2006a).

In addition the Federal Law on Personal Data of 27 July 2006 (Russian Federation 2006b) (updated 2015–2016), defines maintaining the confidentiality of information as an obligatory duty, and requires this information not be transferred to third parties without the direct consent of its owner. According to article 31 of the Fundamentals Of The Legislation Of The Russian Federation On Health Protection No. 5487-1 (2007), "information contained in the person's medical documents shall make up a medical secret."

The situation is also in conflict with the rules of the Declaration of Helsinki, under the heading "Privacy and Confidentiality": "Every precaution must be taken to protect the privacy of research subjects and the confidentiality of their personal information" (WMA 2013: art. 24).

Two other areas of exploitation identified in this proposed clinical trial were the unreasonable exploitation of private time and the financial exploitation of participants/volunteers. The study, as noted above, was very time-consuming for participants and there was no compensation for the expenses of transport, contraceptive products or the disruption of normal daily work and activities. This violates the most recent version of the Declaration of Helsinki: "Appropriate compensation and treatment for subjects who are harmed as a result of participating in research must be ensured" (WMA 2013: art. 15).

This case also points to gender injustice. One could argue that one can detect covert discrimination against vulnerable populations indirectly involved in the study. A fundamental understanding of the gender aspects of research should be guided by the spirit and letter of the Universal Declaration of Human Rights of 1948, which states that "the peoples of the United Nations have in the Charter reaffirmed their faith in fundamental human rights, in the dignity and worth of the human person and in the equal rights of men and women" (UN 1948: preamble).

The ethical conflicts raised by this case study suggest some general arguments that women can be discriminated against through their limited access to participation in clinical trials and the violation of their reproductive rights. The risk of exploitation is especially present when the golden rules of the protection of autonomy, confidentiality and human vulnerability are ignored. The moral force for the realisation of ethical concepts in medical research through the correct process of freely given and obtained informed consent is presented in the UNESCO Universal Declaration on Bioethics and Human Rights (UNESCO 2005) and in many other national and international documents, including the Convention on the Elimination of All Forms of Discrimination against Women, adopted by the United Nations in 1979 (UN 1979).

In summary, the following are the ethical issues raised by this case study:

- gender inequity
- violation of reproductive rights
- inappropriate promises of benefit
- lack of insurance
- confidentiality not preserved
- unreasonable use of private time

Outcome of the Application for Ethics Approval

All properly submitted application documents were reviewed according to the established review procedure (the LEC's standard operating procedure). On the basis of detailed review, discussion took place at a meeting of the LEC with a quorum of its members present. An independent consultant (a specialist in bioethics) was invited to join the meeting after signing an agreement on confidentiality and conflict of interest. Decision-making took place after sufficient time had been allowed for discussion, and was reached by consensus in accordance with the LEC's standard operating procedure. On the basis of disapproving or unfavourable opinions from all members of the LEC, the decision was made in the negative, with detailed and clearly stated reasons provided to the applicant. The clinical trial was not approved.

Lessons Learned and Recommendations

- The system of ethical review worked well in this case, as an unethical study was not approved.
- The possibility of indirectly masking/silencing and blindly exploiting women (pregnant or otherwise) in a study requires attention.
- Gender variety and an assessment of its influence on risk-benefit ratios should be an integral part of clinical trial planning.
- Clinical trials should exclude any opportunity for non-informed or non-agreed interventions that will impact on the privacy of participants' lives, especially in the context of women's reproductive rights.
- Unreasonable risks and burdens, including inadequate compensation and an excessive time burden, must be avoided.

Acknowledgements Thanks to Asmik Asatrian, Dmitrij Ermolenko, Maria Roshina, Elena Briaynceva, Nina Geleznova, Svetlana Ogurcova, and Mariya Ocuneva for their input into the case study.

References

Russian Federation (2005) National standard: Good Clinical Practice. GOST R 52379-2005
Russian Federation (2006a) Federal Law No. 149-FZ of July 27, 2006, on Information, Information Technologies and the Protection of Information (as amended up to Federal Law No. 222-FZ of July 21, 2014). State Duma, Federation Council. http://www.wipo.int/edocs/lexdocs/laws/en/ru/ru126en.pdf
Russian Federation (2006b) Federal Law on Personal Data (as amended by Federal Law of 25.11.2009 No. 266-FZ). State Duma, Federation Council. https://iapp.org/media/pdf/knowledge_center/Russian_Federal_Law_on_Personal_Data.pdf

Russian Federation (2007) Fundamentals of the Legislation of the Russian Federation on Health
    Protection No. 5487-1. https://www.wto.org/english/thewto_e/acc_e/rus_e/WTACCRUS58_
    LEG_270.pdf
Russian Federation (2014) Order of the Ministry of Health No. 125n, 21 March
UN (1948) Universal Declaration of Human Rights. United Nations. http://www.un.org/en/
    universal-declaration-human-rights/
UN (1979) Convention on the Elimination of All Forms of Discrimination against Women.
    Adopted by General Assembly resolution 34/180 of 18 December. http://www.un-documents.
    net/cedaw.htm
UNESCO (2005) Universal Declaration on Bioethics and Human Rights. Adopted by the General
    Conference of the United Nations Educational, Scientific and Cultural Organization, 33rd
    Session, 19 October. http://portal.unesco.org/en/ev.php-URL_ID=31058&URL_DO=DO_
    TOPIC&URL_SECTION=201.html
WMA (2013) WMA Declaration of Helsinki: Ethical principles for medical research involving
    human subjects. World Medical Association. https://www.wma.net/policies-post/wma-
    declaration-of-helsinki-ethical-principles-for-medical-research-involving-human-subjects/

Author Biography

Olga Kubar is head of the Clinical Department at the Pasteur Institute in St Petersburg. She is a
regulatory adviser at national and international level, including UNESCO and the
Interparliamentary Assembly of the Commonwealth of Independent States (IPA CIS). She has
contributed to international guidelines (WHO, UNESCO and IPA CIS) and is a member of the
Russian Bioethics Committee at the Commission of the Russian Federation for UNESCO, and a
former chair and honorary member of the Forum for Ethics Committees in the CIS.

Chapter 8
Healthy Volunteers in Clinical Studies

**Klaus Michael Leisinger, Karin Monika Schmitt
and François Bompart**

Abstract Patients participate in clinical trials for a variety of reasons, the first of which is often the prospect of direct health benefits for themselves. Healthy volunteers, by definition, cannot expect such benefits. In resource-limited settings, healthy volunteers are most often poor people with low literacy levels who might not understand the risks they may be taking and are in no position to refuse financial incentives. For many of them, participation in clinical trials is a critical source of income. An added complication is that some participants covertly enrol in several studies simultaneously, in order to increase their income. This exposes the volunteers to medical risks (e.g. drug-drug interactions), and also potentially biases study data. Our recommendations are that specific efforts are made to ensure proper informed consent of this vulnerable population and that compulsory national databases be established to ensure that healthy volunteers do not participate simultaneously in several studies.

Keywords Clinical trials · Healthy volunteers · Resource-limited settings
Risk · Vulnerability · National databases

Area of Risk of Exploitation

In high-income countries, healthy volunteers are sometimes university students with a good literacy level and reasonable living standards. In resource-limited settings however (including in high income countries), healthy volunteers are most often poor people with low literacy levels who may not understand the risks and are in no position to refuse financial incentives. For many of them, participation in clinical trials is a critical source of income. As a result, even though they might sign informed consent documentation, they are a highly vulnerable group that deserves

K. M. Leisinger (✉) · K. M. Schmitt · F. Bompart
Foundation Global Values Alliance, Schönbeinstrasse 23, 4056 Basel, Switzerland
e-mail: klaus.leisinger@globalvaluesalliance.ch

the "specifically considered protection" recommended by the World Medical Association's Declaration of Helsinki:

> Some groups and individuals are particularly vulnerable and may have an increased likelihood of being wronged or of incurring additional harm.

> All vulnerable groups and individuals should receive specifically considered protection (WMA 2013: art. 19).

The Problem

Informal discussions and a literature review conducted by the authors of this case study have revealed little professional interest in or attention to ethical considerations regarding healthy volunteers from low-income settings. There is very little data published on the number of clinical studies using such volunteers, making it difficult to assess the scope of the issue. While most first-in-human (phase I) clinical trials seem to be performed in high-income countries to ensure the quality of these critical studies, a very large number of studies in healthy volunteers are performed in low- and middle-income countries (LMICs) (Ravinetto 2015:3), particularly bioavailability and bioequivalence studies needed to compare originator and generic medicines.

Clinical studies using healthy volunteers are performed by international as well as local companies, often through contract research organizations (CROs). One of the few papers available on healthy volunteers in LMICs shows how CROs in India resort to "middlemen" to recruit poor participants, who have no understanding of what the studies are about and who sometimes participate in studies without informing their families. They basically "chose to participate in the trials due to insufficient income and unstable jobs" (Krishna and Prasad 2014).

Resource-poor settings are not limited to the LMICs. A few papers describe the situation of healthy volunteers in the US who have become "professional volunteers" and for whom study participation is a way to earn a living (Edelblute and Fisher 2015; Eliott and Abadie 2008). One can assume that many of the ethical issues related to US "professional volunteers" are highly relevant to their counterparts in LMICs. Many have developed tactics to conceal their involvement in several studies at the same time and have become experts at manipulating screening tests for enrolment in clinical trials, for instance by concealing their participation in concomitant studies, medical conditions, concomitant medications or substance abuse (Edelblute and Fisher 2015; Devine et al. 2013). These concealments expose the volunteers to medical risks (e.g. drug-drug interactions) and also potentially bias study data, for instance in terms of safety or pharmacokinetic profiles of the tested drugs (Eliott and Abadie 2008).

The Way Forward

We believe that specific efforts should be expanded to ensure that healthy volunteers are able to understand the key features of the studies (Phase I, II and III) they are offered to participate in, and are therefore able to provide genuine informed consent. This could be done by ensuring that documents are specifically designed for a population with low scientific literacy levels. Establishing compulsory national databases for healthy volunteers appears to be the best way to avoid some of the risks related with participation in multiple studies, detailed above (Devine et al. 2013; Resnik and McCann 2015). Some countries (e.g. France and Morocco) have set up or are in the process of setting up national healthy volunteers' databases to ensure that a given individual's involvement in clinical trials is recorded, that sufficient "wash-out periods" between trials are respected and that payments made to volunteers are tracked so as not to exceed certain levels.

Setting up national databases will require changes in countries' legislation that can only result from the mobilization of key stakeholders, including pharmaceutical companies. In addition to logistical issues that will have to be solved for such systems to be effective, ethical concerns related to confidentiality and data protection issues will have to be addressed. The EU-based pharma industry should support initiatives to ensure that this neglected, highly vulnerable population benefits from the best possible safeguards.

References

Devine EG, Waters ME, Putnam M, Surprise C, O'Malley K, Richambault C, Fishman RL, Knapp CM, Patterson EH, Sarid-Segal O, Streeter C, Colanari L, Ciraulo DA (2013) Concealment and fabrication by experienced research subjects. Clinical Trials 10(6):935–948

Edelblute HB, Fisher JA (2015) The recruitment of normal healthy volunteers: a review of the literature on the use of financial incentives. Journal of Empirical Research on Human Research Ethics 10(1):65–75

Eliott C, Abadie R (2008) Exploiting a research underclass in phase 1 clinical trials. New England Journal of Medicine 358(22):2316–2317

Krishna S, Prasad NP (2014) Ethical issues in recruitment of "healthy volunteers": study of a clinical research organisation in Hyderabad. Indian Journal of Medical Ethics 11(4):228–232

Ravinetto R (2015) Methodological and ethical challenges in non-commercial North-South collaborative clinical trials. Acta Biomedica Lovaniensia 692. KU Leuven, Antwerp

Resnik DB, McCann DJ (2015) Deception by research participants. New England Journal of Medicine 373(13):1192–1193

WMA (2013) WMA Declaration of Helsinki: Ethical principles for medical research involving human subjects. World Medical Association. http://jamanetwork.com/journals/jama/fullarticle/1760318

Author Biographies

Klaus Michael Leisinger founder and president of the foundation Global Values Alliance, is professor of sociology at the University of Basel. He was CEO of the former Ciba Pharmaceuticals regional office in East Africa, and subsequently created and headed the company's philanthropic foundation, serving as president of the Novartis Foundation for Sustainable Development. He served UN Secretary-General Kofi Annan as special adviser on corporate responsibility issues.

Karin Monika Schmitt co-founder and director of the foundation Global Values Alliance, previously shaped the strategic positioning of the Novartis Foundation for Sustainable Development, leading it to consultative status with the United Nations Economic and Social Council. She has directed development projects in Africa, Asia and Latin America.

François Bompart MD, is deputy head and medical director of Sanofi's Access to Medicines Department, with 15 years' experience of clinical trials in resource-limited countries. Since 2012 he has chaired the Global Health Working Group of the European Federation of Pharmaceutical Industries and Associations.

Chapter 9
An International Collaborative Genetic Research Project Conducted in China

Yandong Zhao and Wenxia Zhang

Abstract In 1995, a research team from a renowned US university started collecting blood samples from villagers living in Anhui province, China, with the cooperation of local research institutes and the Chinese government. In 2000, the US university team was accused of violating research ethics principles by not adequately informing the participants about the research and not sharing benefits fairly. Subsequent investigations by American and Chinese media and authorities showed that the US research institute, its research personnel and a pharmaceutical company involved were benefiting substantially from the project, while the Chinese research participants and the government were not. Three levels of exploitation can be distinguished in this case:

- the exploitation of local individual citizens as human research participants
- the exploitation of the local scientific community in China
- the exploitation of the country's national interest

In order to avoid such exploitation, high-income countries as well as low- and middle-income countries should strengthen their institutional arrangements and improve their cooperation mechanisms, in order to ensure that both sides benefit equally from international science and technology cooperation.

Keywords US genetic research team · China · Blood samples
Collaborative study · Exploitation

Y. Zhao (✉) · W. Zhang
Chinese Academy of Science and Technology for Development,
Yuyuantannalu No.8, 10038 Haidian District, Beijing, China
e-mail: zhaoyd@casted.org.cn

© The Author(s) 2018   71
D. Schroeder et al. (eds.), *Ethics Dumping*, SpringerBriefs in Research
and Innovation Governance, https://doi.org/10.1007/978-3-319-64731-9_9

Area of Risk of Exploitation

Genetic studies in urban and rural areas in Anhui province are the topic of this case study. One of the reasons why the case shows a risk of exploitation is that Anhui is not as economically advanced as its neighbouring provinces. For example, in 2015, the GDP per capita in Anhui was CNY 35,997 (EUR 4,878) (Anhui 2016), far lower than that in the more developed neighbouring provinces, such as Jiangsu at CNY 87,995 (EUR 11,925) (Jiangsu 2016), Zhejiang at CNY 77644 (EUR 10,523) (Zhejiang 2016) and Hubei at CNY 50,520 (EUR 6,847) (Hubei 2016).

Background

Since the launch of reform and opening up in the 1970s, international science and technology (S&T) cooperation has been an important means for lifting China's capability and level of S&T innovation. It has also been an indispensable part of China's S&T development. To promote international S&T cooperation, the Chinese government has formulated a series of documents[1] including the *National Outline of International Scientific and Technological Cooperation in the Tenth Five-year Period*; the *Outline for the Implementation of International Scientific and Technological Cooperation Programme in the Eleventh Five-year Period;* and the *Special Programme on International Scientific and Technological Cooperation in the Twelfth Five-year Period.* The national special programme on international S&T cooperation was also added to the system of national S&T programmes in 2001.

In pursuing international S&T cooperation, China has upheld the principles of equality, mutual benefit and common development. From cooperation based on joint research projects in the earlier period to today's all-round cooperation covering skilled professionals, scientific bases and projects, China's international S&T cooperation has grown and continues to grow in both breadth and depth. Through years of development, China has emerged as one of the most important partners for joint scientific research in the world, and has established cooperative relations on S&T with more than a hundred countries and regions. Joint research efforts involving Chinese and international scientific research professionals are growing wider and deeper.

> China's share of global science and engineering publications has pulled within a percentage point of those from the United States, according to the latest research statistics published by the US National Science Foundation (NSF) (Witze 2016).

---

[1]These documents (and others referred to later) are not available in English and have therefore not been included in the reference list.

Of the 82 items in *Top Ten News of Basic Research in China* (later known as the *Top Ten Scientific Advances in China*) between 2005 and 2012, 43 (52% of the total) are about international cooperation projects, while papers based on international cooperation account for 54% of the 100 key academic papers in the relevant fields (Cheng et al. 2015).

Mr Jin Xiaoming, former director-general of the Department of International Cooperation of China's Ministry of Science and Technology, has pointed out that in a world where globalization is the trend in S&T progress, an internationalization strategy is the only way to build China into an innovative country. Without internationalization or international cooperation, China will suffer immensely in its pursuit of advanced S&T (Jin 2012).

For years, international S&T cooperation has played an important role in facilitating China's S&T progress, lifting the scientific research performance and international influence of Chinese scientists, and producing many successful examples of mutually beneficial cooperation. However, it is undeniable that problems of inequality and unfairness also exist in joint research projects, some of which have undermined the interests of the Chinese public and of the scientific community, and even China's national interests.

A strong case in point is that scientific research institutions and personnel from some high-income countries (HICs) have built on their advantages of capital and project experience to make the most of the eagerness of Chinese scientists to make their presence known in the international academic community, and have exploited the flaws and loopholes in China's existing laws and administration to engage in unethical R&D activities in violation of international norms, scientific ethics and even Chinese laws. This has included:

- conducting clinical experiments on human research participants in China which are banned in HICs
- collecting samples in China for commercial purposes
- harvesting China's biological resources and undercutting the intellectual property rights of Chinese scientific research personnel
- conducting human experiments and/or collecting blood samples without providing sufficient information to the participants
- exploiting information asymmetries to conceal information about the experiments
- ignoring and violating the participants' rights to know

These problems are particularly serious in fields that undertake research on medical treatment, pharmacy, genetics, and environmental and air pollution, as well as research projects with potential commercial interests. The "genetic harvest" project conducted by the US University in collaboration with Chinese medical research institutions on farmers in Anhui province in the 1990s is a typical case in point.

Specific Case and Analysis

On 20 December 2000, a *Washington Post* article titled "An isolated region's genetic mother lode" (Pomfret and Nelson 2000) disclosed that a Chinese American researcher of a renowned US University had been collecting blood samples from villagers living in the Dabie Mountains region of China's Anhui province since 1995 with the financial support of the National Institutes of Health (NIH) and biopharmacy companies. The blood samples were transferred to the US university's genetic bank for research into asthma, diabetes, hypertension and other diseases. Because of the value of these carefully selected blood samples to the research and development of new drugs, the US team received a large amount of research funding from international organizations. The report exposed the loss of China's genetic resources and triggered a stir both in China and worldwide.

The US university's genetic harvest project, conducted in Anqing city in Anhui province between 1994 and 1998, involved tens of thousands of farmers in eight counties. The project, led by an associate professor at the US university as the "chief scientist" conducted genetics studies on multiple diseases, including asthma, high blood pressure, obesity, diabetes and osteoporosis, while the experiments on asthma and hypertension were funded by the NIH (Pomfret and Nelson 2000; Xiong and Wang 2001, 2002).

The principal investigator from the US team also collaborated with a US pharmaceuticals company, and received its financial support. The project had three Chinese partners, Beijing Medical University, Anhui Medical University (AMU) and Anqing Municipal Bureau of Public Health. The US-based principal investigator started working with the AMU School of Public Health in 1993, and set up the Anhui Meizhong Bio-medicine and Environmental Health Institute in Anqing. The institute chose the Anqing Bureau of Public Health as its local partner, and selected the population groups suitable for taking samples based on grass-roots investigation. It collected blood samples through physical examination and acquired DNA samples of the target group for research purposes. The joint research project, which was conducted under the guise of free physical examinations for the farmers, mobilized the local population with the help of the local government. Blood samples were collected from farmers in the eight counties of Anqing city: Zongyang, Huaining, Qianshan, Tongcheng, Taihu, Wangjiang, Susong and Yuexi.

Media reports and the complaints of research personnel from the US university later exposed details of certain parts of the project that were suspected of compromising research ethics. The asthma project is an example: the approved number of participants was 2,000, but 16,686 were recruited. The research personnel also changed the amount of the financial subsidies for each recruit for food, travel and job leave allowances; this was intended to be USD 10 per day, but participants were paid an actual amount of only CNY 10 to CNY 20 per day (USD 1.50 to USD 3). In addition, the actual volume of each blood sample was much higher than approved. And the bronchodilators used were also different from what had been approved (Xiong and Wang 2002).

According to the investigation by Chinese journalists, the collection of genetic samples had not been sanctioned by the relevant ethics committee in China (Xiong and Wang 2002). There were also serious breaches of the requirements to keep the participants informed. Many farmers who participated in the physical examination were not aware they were taking part in research. They were never shown or briefed about the "letter of informed consent" , and did not sign or put their fingerprints on any such document. They did not even know which institution they had given their blood samples to, and nobody told them about the real purpose and results of their "physical examination" or the rights and benefits they were entitled to as part of their contribution to research. The asthma project was only one of the dozen human genetic research projects conducted by the US team in China. Other projects also involved the genetic screening of blood samples collected from Chinese farmers for the purpose of establishing the genetic links behind diseases like hypertension, diabetes, obesity and osteoporosis. Many of these projects were first supported by the US pharmaceutical company before NIH funds flowed in (Xiong et al. 2003).

In March 1999, the US University sent a team to China to ensure that the Anhui research was ethically and scientifically sound. Five months later, regulators from the US Department of Health and Human Services launched an investigation into the US university's genetic research in China. In March 2002, the department found that the genetic project in China seriously violated the regulations in multiple respects, including medical ethics, participant safety, and supervision and management (Yangcheng Evening News 2002). On 2 May 2003, the US university published the investigation results of the US government, which stated that there had been some procedural errors in supervision and record-keeping, but no participant was found to have been harmed in any way, so the school would not be penalized (HSPH 2003). Some biomedical experts and ethicists in China expressed regret about these results. They insisted that the studies had apparently violated basic research ethics, and called for a joint US-Chinese review of the experiments (Pomfret and Nelson 2000).

In this international research cooperation on a "genetic harvest", the actors and participants included both international and Chinese research institutes and research personnel, international companies, local government and the local residents who participated in the study.

During this cooperation, the US university, from its commanding position as a world-famous, authoritative international scientific research institute with first-class research personnel and advanced technologies, attracted the participation of Chinese partners and sold them the idea of building partnerships and the opportunity for co-authorship with US research personnel in return for the provision of genetic resources used for research purposes. As a result, they obtained access to a valuable pool of research data resources.

In 2003, the Chinese Ministry of Health and the Chinese Administration of Quality Supervision, Inspection and Quarantine jointly issued regulations limiting the export of special medical articles involving human genetic resources. However, most of the DNA samples the US team had collected in Anhui had already been shipped to the US. The principal investigator himself admitted that for the asthma research alone, 16,400 DNA samples had been transferred to the US (Zhao and Cai 2013). In 2002 and 2003, he set up a biopharmaceutical company and a biopharmaceutical research institute in China. Several Chinese research personnel who had participated in the genetic project in Anhui became his partners.

The US pharmaceutical company became the ultimate beneficiary after supplying research funds. As part of the agreement signed with the US university, they obtained the genetic information of Anhui farmers and claimed that it owned the relevant patents. In July 1995, the company announced that it was in possession of a large collection of asthma genetic samples from China. Soon afterwards, a large Swedish pharmaceutical company, invested USD 53 million in the pharmaceutical company for research into respiratory disease. The company's control of obesity and diabetes genes from China attracted another commitment of USD 70 million from a pharmaceutical giant. The stock price of the company soared from USD 4 per share, when it was listed in May 1995, to more than USD 100 per share in June 2000. Several of the company's senior executives earned a net profit of over USD 10 million each through trade in stocks (Xiong et al. 2003).

In striking contrast, the research participants from China received very few benefits from the project. Chinese research institutes and research personnel did gain the opportunity of working with renowned international research institutes, access to research funds and the co-authorship rights to scientific papers published in international academic journals – all of which appeal to most Chinese scientists – but the local residents who participated in the studies received nothing but a free meal and an insignificant sum of money in travel and job leave allowances. In the words of a Chinese journalist, it was China's national interests and the unprotected Chinese farmers that were most harmed by the project, and it was the big US companies, research institutes and research personnel that received the real benefits (Xiong et al. 2003).

Lessons Learned

This case illustrates the dilemma faced by China and other low- and middle-income countries (LMICs) in conducting international S&T cooperation. On the one hand, opening up and cooperation are an important means for LMICs to build their research capability and achieve faster development. On the other hand, given disadvantages in their capacity for S&T innovation, including the ability to acquire information, and the inadequacy of ethical review and relevant governance systems, it is extremely difficult for them to develop equal partnerships with HICs in S&T cooperation.

Although the academic communities of HICs have established a relatively mature system of ethical standards for scientific research, its research personnel, once working outside their home countries and in a relatively loose regulatory environment, can exploit the systemic loopholes and regulatory "vacuum" of a host country, intentionally or otherwise, and seek improper benefits through potentially illegal acts. In particular, in those research projects driven by commercial interests, when capital uses its economic and technological advantages to exploit resources and benefits from LMICs in the guise of scientific research, those countries find it difficult to resist. Different levels of exploitation might be found in this process, including the exploitation of local individual citizens as human research participants, of the local scientific community and of local countries' national interests.

What has happened in China is something many LMICs have probably experienced. Most of the cases of international institutions, companies and research personnel exploiting China's biological resources happened in the 1980s, the 1990s and the beginning of the 21st century. This had a lot to do with China's inadequate management and regulatory system, the lack of substantive ethical review and insufficient awareness of the need to protect rights and interests during that period. In recent years, the Chinese government and the scientific community have gained a deeper understanding of this problem and have taken a series of positive measures.

In November 1998, the Chinese Ministry of Health established the Committee of Ethical Review on Bio-medical Research Involving the Human Body. To regulate international cooperation in genetics, China promulgated the *Provisional Methods for the Management of Human Genetic Resources* in 1998, which clearly stipulated that international cooperation on China's genetic resources must be conducted on the basis of equality and mutual benefit, with a formal agreement or contract, the approval of the Chinese government and informed consent in the collection of samples.

In 2003, the Chinese Ministry of Health and the Chinese Administration of Quality Supervision, Inspection and Quarantine jointly issued a notice which prescribed that special medical articles involving human genetic resources were not to be taken abroad without authorization. The *Methods for the Ethical Review of Human-involved Bio-medical Research (Provisional)* were promulgated in 2007.

We find that with the gradual improvement of relevant management rules and regulatory systems in China, the number of cases involving the exploitation of China's resources for biological research purposes is diminishing. China has strengthened the rules and regulations concerning intellectual property rights protection, generic resources protection and ethical review in international cooperation, enhanced the relevant management and supervision, and closed the loopholes in the administration of research.

At the same time, as China increasingly opens up to the world, its S&T cooperation with international partners is also widening and deepening, and more and more overseas Chinese students are returning to China. All these factors have greatly mitigated the problem of knowledge and information asymmetry, and have enhanced public awareness of the need to protect rights and interests. As a result, there are fewer cases of HICs using their R&D advantage to exploit China's resources through international cooperation.

In conducting international cooperation, LMICs can only reduce and prevent the occurrence of such cases when they are clear about their own resource advantages, build stronger awareness of the need to protect rights and interests, and improve the relevant management systems. In particular, in terms of the protection and utilization of traditional local knowledge and the protection of rights and interests related to biological resources generally, and genetic resources specifically, LMICs need clear awareness of these issues, while the international community should also give them more protection in this regard.

Recommendations

To reduce exploitation in international S&T cooperation and conduct international cooperation truly on the basis of equality and mutual benefit, we must strengthen our efforts in the following respects:

- We need to foster a stronger awareness of mutually beneficial cooperation in the international community. Countries – big and small, rich and poor, developed and developing – must all uphold the established principles of equality, mutual benefit and sharing in international S&T cooperation, and incorporate these principles into the framework of research ethics and responsible research and innovation.
- The research institutes and research personnel of HICs must abide strictly by the relevant international norms and ethical standards, especially international standards concerning the protection of human rights, and the participants' right to be informed, right to privacy and intellectual property rights. In this context, the regulatory agencies of HICs should strengthen not only the management and supervision of the irregularities happening in their own countries, but also the institutional design, in order to ensure effective supervision of improper acts committed by their research institutes and research personnel in research cooperation with other countries.
- LMICs should strengthen the building of ethical standards, promote knowledge of modern biotechnologies and enhance public awareness of the importance of protecting genetic resources, germplasm resources and patents in order to avoid falling into the trap of technological exploitation, manipulation and deprivation. In particular, they should be alert to the so-called joint R&D activities of certain agencies and the personnel of HICs who might exploit the disadvantages of LMICs and regions – such as poverty, hunger and information asymmetry – in order to use these as the experimental subjects for research and the utilization of technologies without proper compensation.

References

Anhui (2016) Statistical bulletin on national economic and social development in Anhui province in 2015. Anhui Provincial Bureau of Statistics, 25 February [in Chinese]. http://www.ahtjj.gov.cn/tjj/web/info_view.jsp?strId=1461911669310505&_indextow=8

Cheng Y, Liu Y, Wang W (2015) Empirical research on international S&T cooperation promoting the Annual Conference of China Soft Science. Beijing

HSPH (2003) HSPH issues the US government's findings on the school's genetic research in China. Medicine and Philosophy 24(9):46

Hubei (2016) Statistical bulletin on national economic and social development in Hubei province in 2015. Hubei Provincial Bureau of Statistics, 26 February. http://www.stats-hb.gov.cn/tjgb/ndtjgb/hbs/112361.htm

Jiangsu (2016) Statistical bulletin on national economic and social development in Jiangsu province in 2015. Jiangsu Provincial Bureau of Statistics, 29 February. http://www.js.gov.cn/jszfxxgk/tjxx/201602/t20160229492951.html

Jin X (2012) China's internationalization strategy for science and technology and current status of international scientific and technological cooperation [in Chinese]. Science & Technology Industry Parks 11:25–27

Pomfret J, Nelson D (2000) An isolated region's genetic mother lode. Washington Post, 20 December. http://www.washingtonpost.com/wp-dyn/content/article/2008/10/01/AR2008100101158.html

Witze A (2016) Research gets increasingly international. Nature, 19 January. http://www.nature.com/news/research-gets-increasingly-international-1.19198

Xiong L, Wang Y (2001) A suspicious international project of genetic studies. Outlook Weekly 13:24–28

Xiong L, Wang Y (2002) Harvard University's genetic research in China is illegal. Outlook Weekly 15:48–50

Xiong L, Wang Y, Wang C (2003) Poaching China's genetic resources: re-investigating the Harvard genetic project. Outlook Weekly 38:22–25

Yangcheng Evening News (2002) US government: there are serious moral problems in human studies of Harvard. 5 April. http://news.sohu.com/13/95/news148409513.shtml

Zhao X, Cai Z (2013) Social process and development mechanism on biopiracy; case studies from the perspective of constructivism. Studies in Science of Science 31(12)

Zhejiang (2016) Statistical bulletin on national economic and social development in Zhejiang province in 2015. Zhejian Provincial Bureau of Statistics, 24 March. http://www.zj.gov.cn/art/2016/3/24/art_5497_2075286.html

Author Biographies

Yandong Zhao is director of the Institute of Science, Technology and Society of the Chinese Academy of Science and Technology for Development, CASTED, a research arm of the Ministry of Science and Technology.

Wenxia Zhang is a senior researcher at the Institute of Science, Technology and Society of the Chinese Academy of Science and Technology for Development CASTED, a research arm of the Ministry of Science and Technology.

Chapter 10
The Use of Non-human Primates in Research

Kate Chatfield and David Morton

Abstract The use of non-human primates in biomedical research is a contentious issue that raises serious ethical and practical concerns. In the European Union, where regulations on their use are very tight, the number of non-human primates used in research has been in decline over the past decade. However, this decline has been paralleled by an increase in numbers used elsewhere in the world, with less regard for some of the ethical issues (e.g. genetic manipulations). There is evidence that researchers from high-income countries (HICs), where regulations on the use of non-human primates are strict, may be tempted to conduct some of their experiments in countries where regulation is less strict, through new collaborative efforts. In collaborative ventures, equivalence in the application of ethical standards in animal research, regardless of location, is necessary to avoid this exploitation.

Keywords Animal experimentation · Animal ethics · The three 'Rs' Non-human primates

Area of Risk of Exploitation

This case study applies both to academic researchers and to political entities supporting such research. Many areas of research using animals cause public concern, but none more so than those involving non-human primates. European Union Directive 2010/63/EU (EU 2010) imposed several stringent conditions on their use in research, including their acquisition, scientific reasons for their use, husbandry and housing conditions, and record keeping, restricting the overall severity of the procedures carried out, and care of the animals during an experiment. Non-human primates are used in a number of research fields, including neurological research

K. Chatfield (✉)
University of Central Lancashire, Preston PR1 2HE, UK
e-mail: kchatfield@uclan.ac.uk

D. Morton
School of Biosciences, University of Birmingham, Birmingham B15 2TT, UK

© The Author(s) 2018
D. Schroeder et al. (eds.), *Ethics Dumping*, SpringerBriefs in Research and Innovation Governance, https://doi.org/10.1007/978-3-319-64731-9_10

that involves advanced brain responses which can be tracked in various ways, safety testing for novel medicines and new batches of vaccines, defence studies and studies that may benefit wild animals. While in most areas of research the animals concerned might not suffer extremes of pain, in some they are caused significant mental distress.

Certain types of work envisage substantial human benefits (e.g. defence strategies and antidotes), and this may impel some researchers to seek collaboration abroad to carry out work that might be limited or severely curtailed in their own countries. They might also accept compromises in the acquisition of experimental primates: for instance, wild-caught animals, often seen as local pests, could be used instead of purpose-bred animals. Furthermore, the application of the "Three Rs" – replacement, reduction and refinement[1] – is likely to be less stringent, particularly regarding refinement strategies in the housing and husbandry of the animals, and even more so in the experimental design of studies (e.g. the implementation of severity limits and humane endpoints).

Animal Research Worldwide

Animal experimentation is used for many biomedical research activities, including pharmaceutical studies, basic scientific research, biotechnology and traditional medicine research. We cannot determine the exact number of animals used worldwide in research, but there is an estimate of between 50 million and 60 million animal procedures per year, with rats and mice by far the most commonly used species (Understanding Animal Research 2015).

It is estimated that non-human primates represent a very small proportion of the total number of animals used in experiments: fewer than 1 in 1,000 in the EU and approximately 3 in 1,000 in the US (SCHER 2009). Worldwide, however, the number may be more than 100,000 each year.

The wide variety of non-human primate species used in research can be divided into New World species such as marmosets (e.g. the common marmoset, *Callithrix jacchus*), and Old World species such as the long-tailed or cynomolgus or crab-eating macaque (*Macaca fascicularis*) and the rhesus macaque (*Macaca mulatta*). In addition, baboons, another Old World primate of the genus Papio, are occasionally used.

---

[1]The "Three Rs" are the underpinning requirements of most policies and regulations in animal research:

→ Replacement: Methods that avoid or replace the use of animals.
→ Reduction: Methods that minimize the number of animals used per experiment.
→ Refinement: Methods that minimize suffering and improve welfare.

Non-human primates are highly valued in biomedical research because of their genetic similarity to humans,[2] which means they can be especially useful for testing the safety of new drugs and studying infectious diseases, and in neurophysiology, where they can be trained to respond to external stimuli and their central nervous system responses monitored or followed in some way.[3] However, their similarity to humans also raises specific ethical concerns about their use in scientific experiments (SCHER 2009).

In the EU, animal experiments are governed by Directive 2010/63/EU (EU 2010) on the protection of animals used for scientific purposes, which required member states to apply the provisions of the directive through their national legislation from 1 January 2013. According to the directive, the use of non-human primates demands special attention and certain requirements have to be met:

> Due to their genetic proximity to human beings and to their highly developed social skills, the use of non-human primates in scientific procedures raises specific ethical and practical problems in terms of meeting their behavioural, environmental and social needs in a laboratory environment. Furthermore, the use of non-human primates is of the greatest concern to the public (EU 2010: art. 17).

Consequently, the use of non-human primates is strictly controlled and the purposes for which they can be used require rigorous scientific justification:

> Therefore the use of non-human primates should be permitted only in those biomedical areas essential for the benefit of human beings, for which no other alternative replacement methods are yet available. Their use should be permitted only for basic research, the preservation of the respective non-human primate species or when the work, including xenotransplantation, is carried out in relation to potentially life-threatening conditions in humans or in relation to cases having a substantial impact on a person's day-to-day functioning, i.e. debilitating conditions (EU 2010: art. 17).

There are additional requirements on the provision of life histories and severity monitoring that add further criteria to try to ensure that the science is of the highest quality and that animal welfare is not avoidably compromised (EU 2010: art. 30, 39).

With increased scrutiny and regulation, and in response to public opinion, there has been a marked reduction in the number of non-human primates being used in research. Figures show that approximately 6,000 were used in scientific procedures in the EU in 2011, compared with almost 10,000 in 2008 (SCHEER 2016). Furthermore, some institutions are no longer using primates, such as the Harvard Medical School, which closed its affiliated primate facility in 2015. Others are

---

[2]For example, baboons have a 91% DNA similarity (see also Wong 2014).

[3]Safety testing of new drugs, substances and devices, especially those intended for human medicine and dentistry, accounts for approximately 67% of the non-human primates used in research. Fundamental biological research accounts for a further 14% and the research and development of medical and dental products and devices for humans for about 13% (SCHER 2009).

reviewing their primate use: for instance, the US National Institutes of Health announced recently that it would review all non-human primate research that it funds (Cyranoski 2016).

In light of this trend, the European Commission's Scientific Committee on Health, Environmental and Emerging Risks (SCHEER) announced in June 2016 that it was seeking more information to update the EU directive on the use of non-human primate research. In particular, it is seeking opinion on areas of research and testing where non-human primates continue to be used, possibilities to replace their use, and the potential implications for biomedical research, as well as the question of whether the use of non-human primates should be banned altogether in the EU (SCHEER 2016). In Europe, researchers say, the climate for such research is growing colder (Cyranoski 2016).

While the decrease in the number of non-human primates used in the EU may be welcomed and regarded as a beneficial impact of Directive 2010/63/EU, there is rising concern that this decrease has coincided with an increase in the use of non-human primates elsewhere. There is also concern from some in the scientific community that the opportunity to gain valuable insights into certain human diseases will be lost.

Hau et al. (2014) describe how, due to political pressure and the introduction of the new EU directive, biomedical research with non-human primates is increasingly difficult to carry out in Europe. Consequently, European scientists are seeking collaboration with non-human primate centres outside of Europe (Hau et al. 2014).

This has also been noted by Cyranoski (2016), who explains that non-human primate research increasingly faces "a tangle of regulatory hurdles, financial constraints and bioethical opposition" (Cyranoski 2016:300). As a result, some researchers have stopped trying to do such work in the West, and he quotes one neuroscientist as saying that "red tape drove her to China" (Cyranoski 2016).

There is a long tradition of collaboration between European academic institutions and those in the US and Canada, but the network of collaborating institutions is becoming increasingly globalized (Macy 2011). This is highly positive in many respects, but if animals are to be used in collaborative research, the attention to ethical concerns, animal welfare and the quality of the research must be equivalent among research partners around the globe[4] (Bayne et al. 2015). However, regulations, norms, practices and standards in animal research are not currently harmonized, as is clearly illustrated by the following case.

---

[4]The EU have already taken steps towards this end. International projects that are supported by EU funding, such as the Horizon 2020 funding programme, must ensure that all collaborators in the project comply with EU laws in their project activities.

Specific Case and Analysis

In 2013 a report in the British press alleged that an academic from a UK university was bypassing British law in his research with wild-caught baboons in Nairobi (Macrae 2013). A professor of movement neuroscience, part of a team investigating methods to treat conditions affecting the brain such as stroke, spinal cord injury and motor neurone disease, was accused of exploiting a cheap and plentiful source of animals in Nairobi.

The accusation followed an undercover investigation by the British Union for the Abolition of Vivisection (BUAV), which had covertly obtained photos and video footage of the baboons at the relevant institute in Nairobi. BUAV contended that the images revealed disturbing welfare standards and that UK researchers should not accept lower standards when carrying out research at non-UK facilities.

The UK professor was quoted as saying that while animal welfare standards were not as high in Nairobi as in the UK, they had improved greatly during his time there, and that the institute was committed to making further improvements. In addition, he accepted that the experiments would not be permitted in the UK, but argued that it was better to capture wild baboons, who had lived for four or five years in the wild, rather than breed them in captivity. Experiments on wild-caught animals are not normally permitted in the UK, but he claimed that the reasons behind the ban on using wild-caught primates in the UK did not apply to his experiments in Africa.

In a subsequent article in the Kenyan press, the institute in Nairobi denied reports that the facility was being used to conduct harmful research on baboons, claiming that the studies were aimed at advancing medical research for the benefit of Kenya and the world. It added that out of Kenya's 13 non-human primate species, only the two most abundant species (baboons and African green monkeys, another Old World primate) were being used for biomedical research and that, far from being endangered, baboons were considered pests in the wild and those being used in the experiments would otherwise have been killed (Kariuki 2014).

This story received significant coverage in the British media, with celebrities adding their voices to the protests (Nelson 2013). A petition was launched by the students' union at the UK university to persuade the university to end such experiments and, following public pressure, the university decided to halt the baboon experiments in 2014.

There are two immediate concerns that arise from this case: first, that the standards of animal welfare in Kenya may have been lower than the standards required in the EU, and second, that the baboons had been taken out of the wild.

It is not possible to make judgements about the equivalence of standards of animal care without all the facts of the case. However, it is perfectly clear that these experiments would not have been permitted on wild-caught animals in the UK. Of the 2,466 non-human primates used in experiments in the UK in 2014, none had been taken from the wild (Home Office 2015).

It would appear that for many researchers concerns about the equivalence of standards in animal research are fundamental. As Niemi (2011) points out, with an unprecedented level of scrutiny of research possible via the internet, the negative consequences of mere allegations of animal mistreatment are greater than any theoretical advantage to be gained by conducting animal research in a less rigorous environment. This sentiment is echoed by Ogden (2011), who maintains that pharmaceutical and biotech companies do not want to be perceived as using outsourcing in order to bypass standards of humane care and use. Generally, it is acknowledged that the pharmaceutical industry has a vested interest in the promotion of high-quality animal care and facilities and high-quality research outputs (Medina et al. 2015).

However, even for those with the best of intentions, there are challenges for collaborative animal research that stem from a lack of consensus on what should be considered best practice across different cultures. In addition, regulations on animal research and welfare differ from country to country and are subject to change (Landi 2011).

In China, for example, there does not appear to be the same degree of public opposition to the use of non-human primates in research, and many new non-human primate research centres are being established. Some advertise themselves as "primate-research hubs", encouraging researchers to fly in and out and make use of their extensive facilities (Cyranoski 2016).

In Africa, non-human primates are used in research in a number of countries including Kenya, South Africa and Ethiopia. Some Old World primate species and baboons are considered agricultural pests in many parts of Africa, and legislation governing their use in research is generally lacking (Hau et al. 2014).

Most African countries lag behind high-income countries (HICs) in regard to the existence or adequacy of national and/or institutional policies and guidelines on the use of animals in research. While some African countries have been developing ethical or legal frameworks aimed at safeguarding the welfare of animals used for research, in most African countries there is a serious lack of information in the public domain. Consequently, some researchers from (HICs) may be tempted to export their research activities to collaborating African institutions where ethical and legal frameworks on the use of animals may be less stringent (Nyika 2009).

In 2011, Kimwele, Matheka and Ferdowsian published results from their survey of 39 highly ranked academic and research institutions in Kenya aiming to identify those that used animals, their sources of animals, and the application of the Three Rs. At that time, 28 (71.8%) institutions had no designated committee to review or monitor protocols using animals. Only two of the institutions with an established animal care and use committee referred to documented guidelines, and neither documented the composition of their committees (Kimwele et al. 2011).

Across Africa as a whole, the absence of legal and ethical frameworks and committees to review protocols that involve animals in research means that animal protection could be severely compromised, as well as the validity of the scientific outcome data. In addition, the lack of institutional animal ethics committees promotes the outsourcing of animal research to these unregulated institutions (Nyika 2009).

This situation is compounded by the fact that most Western academic institutions do not have much experience with local animal care and use regulations in other countries (Macy 2011). Hence, a double ethics review, where the Western committee also provides an ethics opinion, is not a solution.

Recommendations

- The overarching requirement for avoiding exploitation in animal research is a global code of conduct for research involving animals. There are moves towards this outcome, but it is currently far from resolved. In recent years there have been attempts from different organizations to develop global frameworks. In 2012, the International Council for Laboratory Animal Science and the Council for International Organizations of Medical Sciences updated their International Guiding Principles for Biomedical Research Involving Animals (CIOMS and ICLAS 2012). These principles incorporate the Three Rs and are intended to serve as a framework of responsibility for all countries, including those with emerging research programmes.
- In the absence of a global code of conduct, there will inevitably be variations in standards, regulations, legislation, scientific integrity, data validity and humane practices. In light of this concern, researchers from HICs engaging in collaborative research have an obligation to ensure the application of the same standards that are upheld in their home nations and home institutions.
- For residents of the EU, this entails full compliance with Directive 2010/63/EU (EU 2010) in a manner that is both transparent and auditable. Partner institutions must therefore also be transparent and auditable in the application of principles that are equivalent to those specified in the directive. This must be a requirement even when local legislation and regulation are different or less strict.
- In practice this may entail much closer collaboration with partners on the ground, working together with local representatives to ensure equivalence in all activities such as animal housing and care, as well as experimental procedures.
- European funders of research involving animal experimentation have a particular responsibility to ensure that full compliance with Directive 2010/63/EU is a necessary condition for their support.

Conclusion

Although non-human primates constitute a small proportion of the animals used in research worldwide, their use raises particular ethical concerns. In the absence of a global code of conduct for animal research, animals in countries where regulations and legislation are less well formulated are at risk of exploitation in research. For collaborative ventures, it is vital that institutions from HICs apply precisely the same standards as are required in their home countries and institutions. This may entail close working relationships with local partners to ensure equivalence in standards and some investment to achieve that goal.

For non-human primates, the application of equivalent standards may result in a reduction in the numbers used in collaborative biomedical research, but it will also result in more rigorous science and improved welfare standards and a better application of the Three Rs.

References

Bayne K, Ramachandra GS, Rivera EA, Wang J (2015) The evolution of animal welfare and the 3Rs in Brazil, China, and India. Journal of the American Association for Laboratory Animal Science 54(2):181–191

CIOMS, ICLAS (2012) International guiding principles for biomedical research involving animals. Council for International Organizations of Medical Sciences and International Council for Laboratory Animal Science. http://www.cioms.ch/images/stories/CIOMS/IGP2012.pdf

Cyranoski D (2016) Monkey kingdom: China is positioning itself as a world leader in primate research. Nature 532(7599)

EU (2010) Directive 2010/63/EU of the European Parliament and of the Council of 22 September 2010 on the protection of animals used for scientific purposes (text with EEA relevance). OJ L 276/33. http://eur-lex.europa.eu/legal-content/EN/TXT/PDF/?uri=CELEX:32010L0063&from=EN

Hau AR, Guhad FA, Cooper ME, Farah IO, Souilem O, Hau J (2014) Animal experimentation in Africa: Legislation and guidelines: Prospects for improvement. In: Guillen J (ed) Laboratory animals: Regulations and recommendations for global collaborative research. Elsevier, San Diego CA, p 205–218

Home Office (2015) Annual statistics of scientific procedures on living animals Great Britain 2014. http://www.understandinganimalresearch.org.uk/files/3314/4552/1574/2014_Home_office_animals_stats.pdf

Kariuki J (2014) Institute denies using baboons in harmful research. Daily Nation, 18 August. http://www.nation.co.ke/counties/nairobi/IPR-denies-Goodall-report/1954174-2423042-wb7hrhz/index.html

Kimwele C, Matheka D, Ferdowsian H (2011) A Kenyan perspective on the use of animals in science education and scientific research in Africa and prospects for improvement. Pan African Medical Journal 9(1):45

Landi M (2011) Operational challenges: pharmaceutical industry. In: Animal research in a global environment: meeting the challenges. Proceedings of the November 2008 international workshop. National Academies Press, Washington DC, p 41–45

Macrae F (2013) Outcry as UK scientist flies to Africa for experiments on monkeys that are banned here. Daily Mail, 30 November. http://www.dailymail.co.uk/news/article-2515875/ Outcry-UK-scientist-flies-Africa-experiments-monkeys-banned-here.html

Macy J (2011) Challenges in outsourcing studies: an academic perspective. In: Animal research in a global environment: meeting the challenges. Proceedings of the November 2008 international workshop. National Academies Press, Washington DC, p 232–235

Medina LV, Coenen J, Kastello MD (2015) Special issue: Global 3Rs efforts: making progress and gaining momentum. Journal of the American Association for Laboratory Animal Science 54 (2):115–118

Nelson B (2013) Celebrities speak out against baboon research in Kenya. The Advertiser, 17 December. http://www.durhamadvertiser.co.uk/news/educationzone/news/10882445. Celebrities_speak_out_against_baboon_research_in_Kenya/

Niemi SM (2011) Global issues: operational challenges to working across different standards in academia. In: Animal research in a global environment: meeting the challenges. Proceedings of the November 2008 international workshop. National Academies Press, Washington DC, p 54–61

Nyika A (2009) Animal research ethics in Africa: an overview. Acta Tropica 112(SUPPL. 1): S48–S52

Ogden B (2011) Overcoming challenges: contract research organizations (CROs): setting up a CRO in a foreign country. In: Animal research in a global environment: meeting the challenges. Proceedings of the November 2008 international workshop. National Academies Press, Washington DC, p 46–53

SCHEER (2016) Request for an update to the scientific opinion on the need for non-human primates in biomedical research, production and testing of products and devices. Scientific Committee on Health and environmental Risks. https://ec.europa.eu/health/sites/health/files/ scientific_committees/scheer/docs/scheer_q_001.pdf

SCHER (2009) Non-human primates in research and safety testing. Scientific Committee on Health and Environmental Risks, Health and Consumer Protection Directorate-General, European Commission. http://ec.europa.eu/health/ph_risk/committees/04_scher/docs/scher_o_ 110.pdf

Understanding Animal Research (2015) Numbers of animals. http://www.understandinganimalresearch. org.uk/animals/numbers-animals/

Wong K (2014) Tiny genetic differences between humans and other primates pervade the genome. Scientific American, 1 September. https://www.scientificamerican.com/article/tiny-genetic- differences-between-humans-and-other-primates-pervade-the-genome/

Author Biographies

Kate Chatfield is a social science researcher and ethicist. She is a senior research fellow and Deputy Director of the Centre for Professional Ethics in the School of Health Sciences at the University of Central Lancashire, UK.

David Morton is professor emeritus of biomedical science and ethics at the University of Birmingham. He was, until his retirement in 2012, a member of the European Food Safety Authority scientific panel on animal health and welfare. He was a government adviser on the UK Animals (Scientific Procedures) Act of 1986. He was prime author of the Guidance Document on the Recognition, Assessment, and Use of Clinical Signs as Humane Endpoints for Experimental Animals Used in Safety Evaluation produced by the Organisation for Economic Co-Operation and Development.

Chapter 11
Human Food Trial of a Transgenic Fruit

Jaci van Niekerk and Rachel Wynberg

Abstract The research and development of any "new" agricultural crop created using genetic modification technologies, even if undertaken with the best of intentions, is accompanied by novel human health, environmental, social, economic and other risks. To date, much of the research that has accompanied the release of genetically modified (GM) crops has focused on the environmental and health impacts. Evidence has been inconclusive, however, with debates remaining highly divided and contested, and each "camp" presenting evidence to support its position.

Keywords Genetic modification · Human food trial · Nutritionism

The case presented here does not attempt to elucidate the various positions in the debate, but rather concerns the research process of developing a GM "vitamin-enriched" food for cultivation in a low- or middle-income country (LMIC). It raises questions not only about the ethical complexities of participant involvement in such trials, but also about the ethics of how Northern researchers and philanthropic organizations determine research priorities without necessarily involving the affected LMICs. The case relates to a proposed food trial involving students at a North American university and a banana enriched with a vitamin A precursor, beta-carotene, through genetic modification. The ultimate goal of the study was to roll out the transgenic banana to Uganda, a country where vitamin A deficiency was seen by the researchers and funders involved as a major nutritional challenge.

This case brings up questions about exploitation risks to human participants, and also fuels debates about potential impacts on hunger and nutrition in the intended country of release. By highlighting differences between the concepts of food security and food sovereignty (see table below for definitions), the case illuminates two very different approaches to addressing poverty-induced hunger and malnutrition. Food security, supported by institutions such as the World Bank, the G8-led

J. van Niekerk (✉) · R. Wynberg
Department of Environmental and Geographical Science, University of Cape Town, Cape Town, South Africa
e-mail: jaci.vanniekerk@uct.ac.za

© The Author(s) 2018   91
D. Schroeder et al. (eds.), *Ethics Dumping*, SpringerBriefs in Research and Innovation Governance, https://doi.org/10.1007/978-3-319-64731-9_11

New Alliance for Food Security and Nutrition and large consumer companies, is described by critics as "nutritionism", and is "understood as a set of ideas and practices that seek to end hunger not by directly addressing poverty, but by prioritizing the delivery of individual molecular components of food to those lacking them" (Patel et al. 2015:22). In contrast, food sovereignty aims to reduce malnutrition through an emphasis on diversification and the importance of peoples and countries defining their own food and agricultural priorities, taking into consideration local social, economic, ecological and cultural aspects. Food sovereignty is supported by a growing movement supportive of farmers' rights, women's empowerment and agroecological approaches to farming.

Food security	Food sovereignty
According to the Food and Agriculture Organization of the United Nations (FAO 2001), achieving food security requires: • an abundance of food • access to that food by everyone • nutritional adequacy • food safety	Food sovereignty is defined as "the right of peoples to healthy and culturally appropriate food produced through ecologically sound and sustainable methods, and their right to define their own food and agriculture systems" (Declaration of Nyéléni 2007)

Areas of Risk of Exploitation

This case raises two sets of issues relating to risk of exploitation.

Risks of Participating in the Food Trial

The first set of risks pertained largely to the trial participants in the high-income country (HIC) where the trials were planned. They related to:

- how and whether informed consent was obtained from female students invited to participate in the study
- the potential vulnerability of the student participants, as they may have been unduly incentivized to take part in the study by the USD 900 fee
- potential human health risks, especially given that this was one of the first human food trials of a transgenic plant product.

Risks of Undermining Local Food Systems

The second set of risks pertained to the potential release of the transgenic fruit in Uganda, in the context of pursuing "nutritionism" as a research priority. These include:

- risks of undermining local food and cultural systems and imposing inappropriate solutions
- risks of reducing banana agrobiodiversity

A Proposed Human Food Trial

In early 2014, an US university sent an email to all female students, wanting to recruit 12 volunteers for a transgenic food trial. According to Leys (2014), participants were requested to eat a diet provided by the researchers, which included genetically modified bananas, for four days during each of three study periods. They would have their blood drawn to test whether the fruit's enhanced beta-carotene content translated into higher vitamin A levels in their bodies (AGRA Watch 2016; Gimenez et al. 2016). In return for their participation, the students were offered USD 900. According to a local paper, *The Des Moines Register*, more than 500 applications were received, from whom 12 volunteers were selected (Leys 2015).

The main aim of the study, funded by the Bill and Melinda Gates Foundation (BMGF), was to assess the efficacy of the banana for eventual roll-out in Uganda, an East African country where bananas are a staple food. According to the lead scientist at the US university, the transgenic banana included a gene taken from a sweet variety of banana which naturally produces large amounts of beta-carotene. Residents of Uganda use the less sweet cooking banana as a staple, hence the selection of this type to be genetically modified by the researchers.

Members of non-governmental organizations heard about the proposed study and, along with journalists at *The Des Moines Register*, demanded to know more about it. Initially the lead researcher declined to share information about the study design, claiming that disclosure would be detrimental to her chances of publication (Leys 2014). She relented later, however, stating through a university spokesperson that she had led a similar study five years previously, with six women eating porridge made from corn also modified to produce high levels of beta-carotene (Li et al. 2010).

The proposed food trial, and particularly the initial lack of transparency surrounding it, gained a lot of attention on the US campus, prompting a coalition of concerned students to protest. In collaboration with a number of non-governmental organizations, the students delivered a 57,000-signature petition to the relevant Agriculture and Life Sciences college and to the Gates Foundation's headquarters in February 2016. The students questioned the transparency, risks, and generaliz-ability of the trial, and maintained that prior informed consent had not been obtained from the participants (AGRA Watch 2016). Another exploitation risk hinged on the USD 900 fee, a relatively large sum of money, which could have unduly incentivized students who were not financially secure to take part in the study.

Risks Related to Unknown Human Health Impacts

As this was the first human feeding trial of a GM product that had not been tested on animals, the students were being asked to consume a product of unknown safety (Kruzic et al. 2016). Concerns were also expressed about the potential health risks for women of childbearing age. A molecular biologist based at the Salk Institute for Biological Studies commented:

> Beta-carotene is chemically related to compounds that are known to cause birth defects and other problems in humans at extremely low levels, and these toxic chemicals are possible – if not likely – by-products of plants engineered to make large amounts of beta-carotene. Since there is no required safety testing of the banana or any other genetically modified organism, doing a feeding trial in people, especially women, should not be allowed (AGRA Watch 2016).

Impacts on Local Food and Cultural Systems

Concerns were also raised about the social, economic and environmental impacts of the proposed study. According to Eric Gimenez, executive director of the Institute for Food and Development Policy, such questions

> recognise that hunger and malnutrition are not just biological or technical challenges, they are social problems rooted in poverty, inequality and a skewed distribution of resources. Ending hunger can't be reduced to simple gene transfers, and the socioeconomic and agroecological impacts of GM go far beyond the single crops in which they are genetically expressed (Gimenez et al. 2016).

Such concerns were linked to wider issues about the nutritionism approach adopted through biofortification.[1] Proponents of biofortification recommend its use,

---

[1]Biofortification is the process by which the nutritional quality of food crops is improved through agronomic practices, conventional plant breeding or modern biotechnology [genetic modification]. Biofortification differs from conventional fortification in that biofortification aims to increase nutrient levels in crops during plant growth rather than through manual means during processing of the crops (WHO 2016).

especially in staple crops, to complement conventional fortification[2] activities, particularly in targeting the undernourished in remote rural populations (Bouis et al. 2011; WHO 2016).

Critics of this approach maintain that malnutrition is best countered with a diet based on a diverse variety of foods, and label biofortification as "a strategy that aims to concentrate more nutrients in few staple foods … [and] may contribute to further simplifying diets already overly dependent on a few carbohydrate staples" (Johns and Eyzaguirre 2007:3).

Such views have been supported by the much-studied example from the Philippines and elsewhere of Golden Rice, also fortified with beta-carotene. Stone and Glover (2017), for example, observe that the developers of Golden Rice have yet to produce GM varieties that yield as well as existing varieties, and maintain that the storage qualities of the biofortified rice remains unknown, as does the probability of the beta-carotene being converted into vitamin A in the bodies of severely undernourished children.

Threats to Banana Agrobiodiversity

Aside from questions about the efficacy of the eventual banana product, concerns were also expressed about the risks of undermining local food systems and reducing banana agrobiodiversity. This is especially pertinent considering that East Africa is regarded as a secondary centre of banana diversity (after India), with Uganda being the largest producer and consumer in the region (Gold et al. 2002). If the GM banana variety were to be adopted in Uganda, it would most likely be grown as a monoculture, impacting food security through the erosion of genetic diversity. A significant body of research indicates that diverse plant communities preserve genetic potential for the selection of desirable traits, and also withstand plant pathogens better than monocrops (see e.g. Zhang et al. 2007; Cardinale et al. 2012). Uganda is home to banana varieties that are already higher in beta-carotene than the proposed GM variety. In addition, with the country situated in a fertile, tropical zone, the cultivation of foods naturally rich in beta-carotene, such as sweet potatoes, leafy vegetables and certain types of fruit, offers affordable, healthy, culturally acceptable and locally produced ways of avoiding nutrient deficiency.

---

[2]Fortification is the practice of deliberately increasing the content of an essential micronutrient, i.e. vitamins and minerals (including trace elements), in a food, so as to improve the nutritional quality of the food supply and provide a public health benefit with minimal risk to health (WHO 2016).

Lessons Learned

This case yields a number of valuable lessons across high-, middle- and low-income country scenarios for:

- those involved in plant breeding
- development-related programmes that have LMICs as their focus, as well as funders
- those serving on ethics committees

With regard to the trials conducted in a HIC, the case reveals the importance of ensuring that trial participants make decisions based on a full set of information, and also of examining the rationale behind such trials and fully exploring potential risks to participants. It also suggests that debates about healthy volunteers extend into the domain of agricultural research, a field which is surprisingly undeveloped in the realm of ethics.

With regard to the impacts of such research in LMICs, the case shows that research driven from HICs and by philanthropic donors should be sensitive to local peoples' rights to self-determination of their food systems, and to alternative approaches to addressing nutrient deficiencies. In the words of a concerned Ugandan, "Just because the GM banana has been developed in Australia and is being tested in the US does not make it super! Ugandans know what is super because we have been eating home-grown GM-free bananas for centuries. This GM banana is an insult to our food, to our culture, to us a nation, and we strongly condemn it" (Leys 2015).

Recommendations

- *Human food trials of GM products should be approached with caution.* In the absence of precedents, such trials need to be conducted transparently, with participation based on informed decision-making and without being unduly influenced by financial incentives.
- *The determination of R&D (Research & Development) priorities should be carefully evaluated in terms of local needs, taking into account social, economic, political and environmental implications.* Research involving staple crops should not have outcomes that violate a LMIC's right to self-determination or food sovereignty. Impacts on existing farming systems and on agrobiodiversity need to be carefully considered. Technical solutions such as biofortification should not be introduced at the expense of existing, diverse sources naturally rich in the nutritional substance perceived to be lacking.

- *Further research is needed to deepen understanding about ethics in agricultural research.* This should take cognizance of the need for extensive and inclusive participation in determining research priorities, and should involve regular review to assess the suitability and acceptability of different applications in view of fast-changing technologies.

References

AGRA Watch (2016) Over 50,000 express concern with human feeding trials of GMO bananas. Press release, Community Alliance for Global Justice. http://cagj.org/2016/02/agra-watch-press-release-over-57000-express-concern-with-human-feeding-trials-of-gmo-bananas/

Bouis HE, Hotz C, McClafferty B, Meenakshi JV, Pfeiffer WH (2011) Biofortification: a new tool to reduce micronutrient malnutrition. Food Nutrition Bulletin. 32(I Suppl): S31–S40. doi:10.1177/15648265110321S105

Cardinale BJ, Duffy JE, Gonzalez A, Hooper DU, Perrings C, Venail P, Narwani A, Mace GM, Tilman D, Wardle DA, Kinzig AP, Daily GC, Loreau M, Grace JB, Larigauderie A, Srivastava DS, Naeem S (2012) Biodiversity loss and its impact on humanity. Nature. 486 (7401):59–67. doi:10.1038/nature11148

Declaration of Nyéléni (2007) International Forum on Food Sovereignty, Nyéléni Village, Selingue, Mali. http://viacampesina.org/en/index.php/main-issues-mainmenu-27/food-sovereignty-and-trade-mainmenu-38/262-declaration-of-nyi

FAO (2001) Ethical issues in food and agriculture. Ethical Series No 1. Food and Agriculture Organization of the United Nations, Rome

Gimenez EH, Kruzic A, Carter A, Fidel R (2016) Yes, we need no GMO bananas. The Huffington Post: The Blog, 17 March. http://www.huffingtonpost.com/eric-holt-gimenez/yes-we-need-no-gmo-banana_b_9480400.html

Gold CS, Kiggundu A, Abera AMK, Karamura D (2002) Diversity, distribution and farmer preference of Musa cultivars in Uganda. Experimental Agriculture 38:39–50. doi:10.1017/S0014479702000145

Johns T, Eyzaguirre PB (2007) Biofortification, biodiversity and diet: a search for complementary applications against poverty and malnutrition. Food Policy 32(1):1–24

Kruzic A, Carter A, Fidel R (2016) When a banana is much more than a banana. Food First, 7 March. https://foodfirst.org/when-a-banana-is-much-more-than-a-banana/

Leys, T (2014) ISU researcher to test altered bananas. The Des Moines Register, 2 August. http://www.desmoinesregister.com/story/news/2014/08/02/isu-researcher-test-genetically-altered-bananas/13502003/

Leys, T (2015) Iowa trial of GM bananas is delayed. The Des Moines Register, 12 January. http://www.desmoinesregister.com/story/news/health/2015/01/12/isu-genetically-modified-bananas-trial/21663557/

Li S, Nugroho A, Rocheford T, White WS (2010) Vitamin A equivalence of the β-carotene in β-carotene-biofortified maize porridge consumed by women. The American Journal of Clinical Nutrition 92(5):1105–1112. doi:10.3945/ajcn.2010.29802

Patel R, Bezner Kerr R, Shumba L, Dakishoni L (2015) Cook, eat, man, woman: understanding the New Alliance for Food Security and Nutrition, nutritionism and its alternatives from Malawi. The Journal of Peasant Studies 42(1):21–44. doi:10.1080/03066150.2014.971767

Stone GD, Glover D (2017) Disembedding grain: Golden Rice, the green revolution, and heirloom seeds in the Philippines. Agriculture and Human Values 34(1):87–102. doi:10.1007/s10460-016-9696-1

WHO (2016) Biofortification of staple crops. World Health Organization, e-Library of Evidence for Nutrition Actions (eLENA). http://www.who.int/elena/titles/biofortification/en/

Zhang W, Ricketts TH, Kremen C, Carney K, Swinton SM (2007) Ecosystem services and dis-services to agriculture. Ecological Economics. 64:253–260

Author Biographies

Jaci van Niekerk is a researcher in the Department of Environmental and Geographic Science at the University of Cape Town.

Rachel Wynberg holds a South African Research Chair on Environmental and Social Dimensions of the Bio-economy, based in the Department of Environmental and Geographical Science, University of Cape Town. Over the past 20 years she has advised governments, industry, civil society organizations and international agencies and is actively involved with NGOs in southern Africa.

Chapter 12
ICT and Mobile Data for Health Research

David Coles, Jane Wathuta and Pamela Andanda

Abstract Mobile cellular subscriptions had reached 87% of the world's population by 2011 (ITU 2011). Notably, Africa has "the fastest mobile phone growth rate in the world and … a proliferation of social media users" (Mutula in Information ethics in Africa: cross-cutting themes. African Centre of Excellence for Information Ethics, Pretoria, pp 29–42, 2013:31). Mobile phones that can run software applications (apps) are increasingly used in health settings, for example, to improve diagnosis and personalize health care (Mosa et al. in BMC Medical Informatics and Decision Making 12(1):67, 2012). This fast-paced development saw the number of "mHealth" apps reach 97,000 as of March 2013 (He et al. in AMIA Annual Symposium Proceedings, pp 645–654, 2014).

Keywords Health research · ICT · mHealth · Mobile data · Mobile phones
Personal data

The application of mobile technologies (mobile phones or other remote monitoring devices) for health-related purposes is termed "mHealth": a mobile tool for expanding access to health information and services around the world (K4Health 2014). According to the World Health Organization (WHO 2011:6) , mHealth is the "medical and public health practice supported by mobile devices, such as mobile phones, patient monitoring devices, personal digital assistants and other wireless devices". Although mHealth has come to signify the use of any mobile technology to address health care challenges such as access, quality, affordability, matching of resources and behavioural norms (Qiang et al. 2011), most mHealth interventions use mobile phone technology, thanks to its versatility as an ICT tool (Leon and Schneider 2012:7).

D. Coles (✉)
University of Central Lancashire, Centre for Professional Ethics, The Bacchus House, Elsdon,
Newcastle upon Tyne NE19 1AA, UK
e-mail: david.coles@hazyrays.com

J. Wathuta · P. Andanda
University of the Witwatersrand, Private Bag 3 Wits, Johannesburg 2050, South Africa

© The Author(s) 2018
D. Schroeder et al. (eds.), *Ethics Dumping*, SpringerBriefs in Research
and Innovation Governance, https://doi.org/10.1007/978-3-319-64731-9_12

With the pervasive growth in technology infrastructure, mHealth can reach communities in ways that conventional health services and other communication tools cannot. Mobile phones are described as potentially the most widespread embedded surveillance tools, especially due to the use of location sensors and the consequent possibility of documenting and quantifying habits, routines, and personal associations (Shilton 2009). This case study focuses on the potential ethical issues associated with the use of mHealth apps in medical research and health care. mHealth offers "attractive low-cost, real-time ways to assess disease, movement, images, behaviour, social interactions, environmental toxins, metabolites" (Collins 2012:1). It has the power to bring the research lab to the patient and obtain real-time, continuous biological, behavioural and environmental data (Collins 2012).

Mobile phones collect a wide range of personal information from their users, with or without their knowledge, which raises novel and complex ethical and practical challenges. Research teams (and clinicians) need to understand these challenges so that, without rejecting mHealth and related mobile technological advancements, they minimize any unintended harms (Carter et al. 2015). Wicklund (2015) observes that clinical studies that utilize mHealth devices and platforms are venturing into uncharted ethical territory.

Area of Risk of Exploitation

Software apps in the mHealth category can be used for collecting health-related data on a large scale for biomedical research; the so-called "big data" (Park and Jayaraman 2014; Hsieh et al. 2013). In general, however, mHealth raises concerns regarding data security issues – from transmission of data to its local storage, and "ownership" of what is otherwise considered confidential patient data. This data is easy to obtain, but difficult or impossible to retract once shared. In addition to safety and security risks, mobile sensing also disrupts social boundaries and challenges distinctions between public and private (Shilton 2009). One of the key challenges of using mHealth in low- and middle-income countries (LMICs) is how to ensure workable approaches to privacy and security (Leon and Schneider 2012:19).

Carter et al. (2015) have identified a range of ethical issues raised by the use of mobile phones for research and clinical purposes. These are:

- the protection of privacy
- minimizing third-party uses of data
- informing patients of complex risks when obtaining consent
- maximizing benefits while minimizing the potential for disclosure to third parties
- care in the communication of clinically relevant information
- the rigorous evaluation and regulation of mHealth products before widespread use

In practical terms, the issues discussed below need to be considered carefully.

Context-Based and Fully Informed Consent Should Be Obtained

Researchers should seek and obtain informed consent before using mHealth technologies in research. Accordingly, participants must be informed about, and understand the risks and benefits of, mHealth technologies, and then make a free and voluntary decision to participate or not. The risks associated with mHealth are complex, and these need to be communicated and negotiated. If the study involves the collection of data from interaction with identifiable third parties, it may be necessary to obtain their informed consent as well. This in turn means that mHealth participants will have to disclose their condition and/or mHealth participation (Carter et al. 2015).

Only Necessary Data Should Be Collected

Compared with other health information systems, mHealth collects a much larger amount and broader range of data about patient lifestyles and activities, over an extended period (He et al. 2014). A potential danger to bear in mind in this regard is that of collecting excessive amounts of raw data to maximize the information extracted by the research team (Carter et al. 2015).

Any Tracking Should Be Proportionate and the Correct Person Should Be Tracked

Continuous or intermittent recording and transmission of detailed information about where a person is, and to some extent what they are doing, may breach privacy and confidentiality. There are risks of inadvertent insight into a participant's behaviour revealing information beyond the profiles that are scientifically justified and for which data collection was employed. This also poses problems of informed consent, as privacy may be violated in ways unforeseen by either investigators or participants. Text messages (SMS) can be read by persons other than the intended recipient; messages can be forwarded and can remain on unsecured devices indefinitely. One result could be the unintended disclosure of a medical condition (Labrique et al. 2013).

Research Participants Should Know Exactly Which Data Is Collected and Who Will or Could Have Access to It

This is a great challenge, especially in a global research environment that increasingly requires the sharing of data in publicly available repositories. The case of an alleged breach of smartphone users' privacy by manufacturers of popular smartphone apps for Apple and Android devices illustrates this risk. The manufacturers are alleged to have gathered information from personal address books on the phones of Kenyan users, stored it on their own computers, and transmitted it without the knowledge of its owners, all of which demonstrates how difficult it is to guarantee privacy when using smartphones (Mutula 2013; Wambugu 2012).

The security of data collected via mobile phones cannot be guaranteed either, in part because no strict privacy regulations exist.[1] Many mHealth apps do not use encryption when transferring data, and even when they do, hackers and governments can still gain access. Potential violations of privacy include hacking of personal data with the known likelihood of identity theft and financial losses, computer malware and virus programs, and malevolent apps planted by developers who steal data for commercial or criminal purposes (He et al. 2014).

Incentives to Take Part in Research Should Be Proportionate and not Result in Exploitation

Research involving mHealth apps often requires the participant to have a smartphone. If researchers specifically target those who do not already own newer devices or other modes of mobile technology, the prospect of being given access to such technology may unduly influence them to take part (Labrique et al. 2013:3). Patients should not, however, be excluded from mHealth monitoring benefits if they cannot afford a device capable of supporting the app or connect with networks capable of transmitting potentially large volumes of data. This requirement therefore needs very careful judgement.

---

[1]Companies like Apple and Google have to comply with the privacy regulations in each of the countries where they collect data. Where little or no privacy regulation exists, the companies have wide scope regarding what data they collect and how they use it. Interestingly, Apple announced that with their new iOS10 operating system they would be introducing "differential privacy", which they claimed would enable them to collect much more personal user data while preserving users' privacy. This concept involves introducing numerical "noise" into the data collected in order to de-identify it (see Brandom 2016). However, it is questionable whether data provided this way will be suitable for research purposes (see Friedman and Schuster 2010).

Specific Case and Analysis

The details of a case of HIV/AIDS tele-counselling in South Africa were obtained from an interview with Cell-Life's general manager, Peter Benjamin, conducted and published in 2011 by Boyle (2011). Additional information is available in a report that was prepared on the use of mobile technologies for the monitoring and evaluation of public sector community-based health services (Leon and Schneider 2012).[2]

Cell-Life, a non-profit organization, entered into a contract with the South African national Department of Health (DOH) for a big project. "Cell-Life started in 2001 as a research collaboration between staff of the engineering faculty of the University of Cape Town (UCT) and the Cape Peninsula University of Technology (CPUT)" (Loudon and Rivett 2013). It became a not-for-profit organization in 2006 (Loudon and Rivett 2013). In terms of the contract, the DOH set up a national mHealth system that used cellphones for monitoring an HIV counselling and testing (HCT) campaign.

Cell-Life used chat software called Mxit, which enabled users to send instant messages over a cellphone system. To do this, users had to download a small app that connected them to the Mxit server, enabling immediate communication with anyone else on Mxit. The app sent SMS-type messages through GPRS,[3] via which messaging was effectively free.

Cell-Life created a website within Mxit where it provided all the usual HIV content, information and interactive quizzes. An interesting feature that Cell-Life included was linking Mxit to South Africa's National AIDS Helpline, so that users could text on Mxit and the message would go through to the computer screen of a professional HIV counsellor at the National AIDS Helpline. The counsellor would type a reply which would appear on the user's cellphone screen.

Cell-Life was awarded additional contracts by the DOH for the design and implementation of a mobile monitoring and reporting system for the national HIV counselling and testing (HCT) campaign, and the national antiretroviral treatment expansion programme (Cell-Life nd). These systems have been the subject of research into how software applications for the monitoring and evaluation of community-based care are used in a research and service delivery context (Leon and Schneider 2012).

The data processed and transmitted through the software apps related to patients' personal information, which was subsequently stored and monitored through the system. The use of mobile phones in this process raises practical ethical issues, such as concerns about the protection of information and privacy, and consent to the potential use of such information for research purposes. As Labrique et al. (2013)

---

[2]See also Cell-Life (nd).

[3]"General Packet Radio Service (GPRS) is a packet oriented mobile data service on the 2G and 3G cellular communication system's global system for mobile communications (GSM)" (General Packet Radio Service 2017).

have observed, although mHealth apps ensure the availability of real-time data that brings with it new and beneficial strategies, the rapid adoption of these technologies raises ethical issues that need careful consideration. Accordingly, existing standards and practices have to be supplemented with new guidelines to ensure that patients and vulnerable populations are adequately protected. The gap between technological innovation and the development of ethical standards and guidance needs to be reduced, so that researchers and other stakeholders have a reference framework for assessing and mitigating the risks of mHealth research and data collection.

Recommendations

The following measures could help avert the possibility of exploitation in the context of mHealth:

- Developers should determine when, where and how sensitive data are uploaded and stored, to minimize the risk of privacy violations. In addition, they should take steps, by using encryption and anonymization (Carter et al. 2015; He et al. 2014), to ensure that data collected by an mHealth app are not available to other apps or programs installed on the phone or in third-party storage without security and privacy guarantees (He et al. 2014).
- Participants should be able to control what they consent to and how their data may be used and stored. The data should be deleted as soon as no longer needed (Albrecht and Fangerau 2015).
- Appropriate regulation of mHealth devices and apps should be developed to ensure their safety and effectiveness, including minimal privacy violations and guarantees that they provide clinically accurate information. Albrecht and Fangerau (2015), for instance, have recommended the transformation of the fundamental principles of medical ethics in order to make them applicable to mHealth.
- Proven innovations for the improvement of data protection and privacy should be implemented by researchers as soon as possible after they become available.

References

Albrecht U-V, Fangerau H (2015) Do ethics need to be adapted to mHealth? Studies in Health Technology and Informatics 213:219–222

Boyle C (nd) (2011) mHealth benefits: little evidence – yet (2011) http://www.mobilethinkers.com/2011/01/mhealth-benefits-no-evidence-%E2%80%93-yet/

Brandom R (2016) This is what Apple's differential privacy means for iOS 10. The Verge, 17 June. http://www.theverge.com/2016/6/17/11957782/apple-differential-privacy-ios-10-wwdc-2016

Carter A, Liddle J, Hall W, Chenery H (2015) Mobile phones in research and treatment: ethical guidelines and future directions. JMIR mHealth and uHealth 3(4): e95

Cell-Life (nd) Mobile ME for the national HIV counselling and testing campaign. http://www.cell-life.org/projects/health-care-and-testing/

Collins F (2012). The real promise of mobile health apps: mobile devices have the potential to become powerful medical tools. Scientific American 307(1):1

Friedman A, Schuster A (2010) Data mining with differential privacy. In: Proceedings of the 16th ACM SIGKDD international conference on knowledge discovery and data mining. Association for Computing Machinery, New York, p 493–502

General Packet Radio Service (2017) Wikipedia: The Free Encyclopedia. Wikimedia Foundation Inc. https://en.wikipedia.org/wiki/General_Packet_Radio_Service

He D, Naveed M, Gunter CA, Nahrstedt K (2014) Security concerns in Android mHealth apps. AMIA Annual Symposium Proceedings 645–654

Hsieh J-C, Li A-H, Yang C-C (2013) Mobile, cloud, and big data computing: contributions, challenges, and new directions in telecardiology. International Journal of Environmental Research and Public Health 10(11):6131–6153

ITU (2011) The world in 2011: ICT facts and figures. International Telecommunication Union. http://www.itu.int/ITU-D/ict/facts/2011/material/ICTFactsFigures2011.pdf

K4Health (2014) The mHealth planning guide: key considerations for integrating mobile technology into health programs. https://www.k4health.org/toolkits/mhealth-planning-guide

Labrique AB, Kirk GD, Westergaard RP, Merritt MW (2013) Ethical issues in mHealth research involving persons living with HIV/AIDS and substance abuse. AIDS Research and Treatment 2013:189645 1–6

Leon N, Schneider H (2012) MHealth4CBS in South Africa: a review of the role of mobile phone technology for monitoring and evaluation of community based health services. South African Medical Research Council, University of the Western Cape. http://www.mrc.ac.za/healthsystems/MHealth4CBSReview.pdf

Loudon M, Rivett U (2013) Enacting openness in ICT4D research. In: Smith LM, Reilly MAK (eds) Open development: networked innovations in international development. MIT Press, Cambridge MA, p 53–78

Mosa ASM, Yoo I, Sheets L (2012) A systematic review of healthcare applications for smartphones. BMC Medical Informatics and Decision Making 12(1):67

Mutula SM (2013) Ethical dimensions of the information society: implications for Africa. In: Ocholla D, Britz J, Capurro R, Bester C (eds) Information ethics in Africa: cross-cutting themes. African Centre of Excellence for Information Ethics, Pretoria, p 29–42. http://www.africainfoethics.org/pdf/ie_africa/chapter_4.pdf

Park S, Jayaraman S (2014) A transdisciplinary approach to wearables, big data and quality of life. 36th annual international conference of the IEEE engineering in medicine and biology Society, Chicago IL, p 4155–4158

Qiang CZ, Yamamichi M, Hausman V, ltman D (2011) Mobile applications for the health sector. World Bank, Washington DC

Shilton K (2009) Four billion little brothers? Privacy, mobile phones, and ubiquitous data collection. Communications of the ACM 52(11):48–53

Wambugu S (2012) Cellphone giving away your personal privacy. Daily Nation, 18 February. http://www.nation.co.ke/oped/Opinion/Cellphone-giving-away-your-personal-privacy-/440808-1330412-tdkawuz/index.html

WHO (2011) mHealth: new horizons for health through mobile technologies: second global survey on eHealth. World Health Organization, Geneva. www.who.int/goe/publications/goe_mhealth_web.pdf

Wicklund E (2015) New project to create mHealth ethics for clinical trials. mHealth Intelligence, I December. http://mhealthintelligence.com/news/new-project-to-create-mhealth-ethics-for-clinical-trials

Author Biographies

David Coles is senior research fellow at the University of Central Lancashire's Centre for Professional Ethics and a research associate at the University of Newcastle's School of Agriculture, Food and Rural Development. Previously he was joint programme manager for the European and Developing Countries Clinical Trials Partnership, as well as developing and implementing the EU system of ethics review for the EU Framework Programme.

Jane Wathuta is a postdoctoral research fellow at the School of Law, University of the Witwatersrand, Johannesburg. She previously worked with Strathmore University and the Kianda Foundation educational trust in Nairobi. She is an advocate of the High Court of Kenya and a member of the Research Ethics Committee of the Centre for Research in Therapeutic Sciences, the Kenya Medical Research Institute, South Africa's Council for Scientific and Industrial Research and the African Centre for Clinical Trials.

Pamela Andanda is a professor of law at the University of the Witwatersrand, Johannesburg. Pamela is a member of UNESCO's International Bioethics Committee, the Ethics Review Committee of Strathmore University's Center for Research in Therapeutic Sciences, and the Data and Biospecimen Access Committee of the Human Heredity and Health in Africa (H3Africa) Consortium.

Chapter 13
Safety and Security Risks of CRISPR/Cas9

Johannes Rath

Abstract This case study looks into recent developments with regard to the CRISPR/Cas9 and other novel genome editing technologies that are becoming widely available thanks to their low costs and modest technological requirements.

Keywords Biosafety · Biosecurity · CRISPR/Cas9
Responsible research and innovation

Genome editing allows the specific modification of a genome; genes are modified within their respective location in the genome, making the changes often indistinguishable from natural mutations. Developments of this technology such as the use of gene drives, where specific genes are spread within populations, or the use of viral vector systems, are enabling additional applications in environmental engineering and disease treatment. There are substantial individual and societal benefits from applying genome editing; nonetheless the technology also poses significant risks to individuals, society as a whole and the environment.

The central focus of this case study is on the unresolved ethical issues related to safety and security that pose both short-term and long-term challenges to international research partnerships. As such, the case study focuses not on a single incident but on the risks in the proliferation of a new and very powerful technology at a time when accepted and tailored ethical and legal frameworks at the international, national and local level are missing.

In the case study two areas of *safety* risks are mapped and existing governance approaches described: first, risks to humans, for example in relation to therapeutic applications of genome editing; second, risks to the environment in relation to the use of genome editing on animals, plants and microbes. In addition, two aspects of *security* risks are also assessed: first, the creation of harmful agents relevant in the

J. Rath (✉)
Department Integrative Zoologie, University of Vienna, Althanstrasse 14, 1090 Vienna, Austria
e-mail: johannes.rath@univie.ac.at

© The Author(s) 2018
D. Schroeder et al. (eds.), *Ethics Dumping*, SpringerBriefs in Research and Innovation Governance, https://doi.org/10.1007/978-3-319-64731-9_13

bioweapons context; second, human enhancement in a military context and its medium- and long-term implications for international security.

It is concluded that the rapid emergence of high-risk safety and security applications of genome editing challenge not only today's safety and security risk assessment but also existing governance tools. In addition, the absence of international standards of governance may result in safety- and security-sensitive experiments being transferred to countries with less stringent oversight, which will have serious implications for trust in international research.

Area of Risk of Exploitation

The key area of risk relates to the exploitation of international inconsistencies in biosafety and biosecurity with regard to the governance of genome editing experiments. These inconsistencies create an environment where risky experiments might be carried out in countries with no legal framework (European Commission nd), or in countries where, although legal frameworks exist, their implementation cannot be achieved due to limited resources (Dickmann et al. 2015). This undercuts established European standards of safety and security, while at the same time, due to the nature of some of these experiments, potentially affecting safety and security in Europe itself (Defensive Drives 2015).

Analysis

In everyday life, the terms "safety" and "security" are often used interchangeably. Here "safety" denotes the protection of humans, animals, plants and the environment from unintentional harm, whereas "security" relates to intentional harm (e.g. in a military context). This case study on genome editing focuses on the safety and security implications in four concrete experimental settings that have either been used in laboratories already, or are well within the range of existing technological capacities. These experimental settings are:

- the use of genome editing in human inheritable disease, infectious disease and cancer treatment and human enhancement
- the use of genome editing in creating novel pathogenic organisms
- the use of genome editing in environmental engineering and disease vector eradication
- the use of genome editing in agriculture

The controversy surrounding the publication of a research paper applying genome editing technologies to human embryonic stem cells has brought to the attention of the international scientific community the varying international governance approaches regarding such research. Since then a broad discussion has emerged on how to use this technology in an ethically sound way (Cyranoski 2015:272; Lanphier and Urnov 2015:411; Callaway 2016:16).

Although many of these discussions focus on the moral status of a human embryo and the permissiveness of human germ-line enhancement, it has become generally accepted that a common ethical issue is whether or not genome editing can be carried out safely and securely.

The safety aspect was highlighted very early on in the discussion as a critical limitation that would need to be resolved before any application of genome editing on humans or release into the environment could take place (Akbari et al. 2015). The security aspect, on the other hand, only recently gained attention when leading governmental officials identified genome editing as a national security threat (Oye et al. 2014).

Resolving the major safety and security concerns of genome editing is therefore of general importance, not only as a prerequisite for a reasonable discussion of the potential benefits, but also to foster trust among stakeholders in international collaborative research.

Genome editing has huge potential in human inheritable disease treatment and human enhancement. Research here relates to the treatment of various genetic disorders, infectious diseases and cancer. Recent examples that are currently undergoing safety testing in clinical trials are the use of somatic gene therapies involving immune cell modifications to treat cancer (Reardon 2016), CRISPR-based approaches to treating HIV (Reardon 2014) and the proof of principle of genome editing in the treatment of heritable diseases such as Duchenne muscular dystrophy (Mendell and Rodino-Klapac 2016). Key safety concerns in this area have been the number of off-target changes, mosaicism and potential epigenetic effects (Next-generation genome editing 2015). These are not new safety concerns, but have also been encountered in other gene therapeutic approaches. The existing step-wise approach applied in clinical studies should therefore be sufficiently robust to identify, assess and govern such risks.

There is a fluid relationship between genome editing as employed in heritable disease treatment and its use for human enhancement (Ishii 2015; Cox et al. 2015). Genetic human enhancement has substantial security implications. In certain countries, approving the use of genome editing for this purpose (e.g. IQ and physical endurance) would have far-reaching military and economic security implications at the national and international level. These security risks need to be included in risk benefit assessments of human enhancement based on genome editing.

Certain genome editing techniques open the possibility for the development of a new class of infectious pathogenic organisms. A recent example has been the creation of cancer models in mice, where the cancerous mutation was introduced through genome editing using viral vectors – in essence transforming cancer into a transmissible infectious disease (Chiou et al. 2015). This creates novel safety risks that will need to be included in biosafety oversight schemes. In addition, such work has the potential to create new generations of biological and chemical weapons which might not be detectable by current diagnostics.

The use of genome editing in environmental engineering has been discussed in the context of pest control, with new ways to eradicate agricultural pests (Huang et al. 2016; Leftwich et al. 2016), as well as that of disease eradication. For example, gene drive systems are being developed to eradicate malaria (Gantza et al. 2015), and contemplated for the eradication of the Zika (Hegg 2016) arthropod vector. Key safety concerns relate to the environmental harmfulness, controllability and reversibility of such environmental interventions. Key security concerns relate to their potential use as socio-economic and environmental weapons.

The use of gene drives in an environmental context creates novel risks for both safety and security, which are not restricted by national boundaries. Current national and international risk management approaches to biosafety and biosecurity are incapable of mitigating these risks adequately.

The use of genome editing in agriculture for breeding purposes in plants and animals (Sovová et al. 2016) creates unique and novel challenges to biosafety and biosecurity. Key safety concerns relate to the outbreeding and spread of these new varieties into natural populations, the detectability of these new variants (Breeding Controls 2016) and challenges to established coexistence provisions (Ledford 2015).

Below are quotations from leading researchers that address some of the relevant issues on biosafety and biosecurity (all quoted in Ledford 2015):

Leading Researchers	Quotes
James Haber – on the issue of off-target effects:	These enzymes will cut in places other than the places you have designed them to cut, and that has lots of implications.
Jennifer Doudna – on the biosafety and biosecurity of an experiment creating a human cancer model through a CRISPR-engineered virus:	It seemed incredibly scary that you might have students who were working with such a thing. ... It's important for people to appreciate what this technology can do.
George Church – on the safety risks of gene drives in relation to the environment:	It has to have a fairly high pay-off, because it has a risk of irreversibility – and unintended or hard-to-calculate consequences for other species.
Jennifer Kuzma – on the detectability of genome-edited GMOs in nature:	With gene editing, there's no longer the ability to really track engineered products. It will be hard to detect whether something has been mutated conventionally or genetically engineered.
Kenneth Oye – on governance:	It is essential that national regulatory authorities and international organizations get on top of this — really get on top of it.

Recommendations

There are four levels on which recommendations can be made to avoid the exploitation of safety and security weaknesses in genome editing in the future.

Technical Level

- Reduce off-target effects, mosaicism and epigenetic effects through further research in higher fidelity and better understanding of genome editing technologies.
- Use safe virus systems or alternative less risky vector systems to transfer genome editing tools.
- Develop reversal gene drives in parallel that can undo the effects of gene drives.
- Provide technological assistance (e.g. detection capacities for modified organisms) in implementing international obligations such as the Cartagena Protocol.

Containment Level

- Ensure adequate biosafety risk classification and implementation of adequate containment measures in biosafety-sensitive genome editing experiments.
- Develop "molecular containment" approaches when working with genome-edited high-risk pathogens.

Governance and Oversight level

- Provide international guidance or amend existing guidance documents on biosafety and biosecurity to cover risks from genome editing.
- Map the status of existing biosafety and biosecurity legislation as well as its practical implementation in countries carrying out genome editing experiments.
- Include stakeholders (e.g. funding institutions, research institutions, researchers) in the responsible governance of research involving genome editing.

International Standardization

- In case of gaps in legal oversight, develop international codes and guidelines for safe and secure work in genome editing.

References

Akbari O, Bellen H, Bier E, Bullock SL, Burt A, Church GM, Cook KR, Duchek P8, Edwards OR, Esvelt KM1, Gantz VM, Golic KG, Gratz SJ, Harrison MM, Hayes KR, James AA, Kaufman TC, Knoblich J, Malik HS, Matthews KA, O'Connor-Giles KM, Parks AL, Perrimon N, Port F, Russell S, Ueda R, Wildonger J (2015) Safeguarding gene drive experiments in the laboratory. Science 349(6251):927–929. doi: 10.1126/science.aac7932

Breeding controls (2016) Editorial. Nature 532(7598):147. doi: 10.1038/532147a

Callaway E (2016) Embryo editing gets green light. Nature 530:16

Chiou SH, Winters IP, Wang J, Naranjo S, Dudgeon C, Tamburini FB, Brady JJ, Yang D, Grüner BM, Chuang CH, Caswell DR, Zeng H, Chu P, Kim GE, Carpizo DR, Kim SK, Winslow MM (2015) Pancreatic cancer modelling using retrograde viral vector delivery and in vivo CRISPR/Cas9-mediated somatic genome editing. Genes & Development 29(14):1576–1585. doi: 10.1101/gad.264861.115

Cox DBT, Platt RJ, Zhang F (2015) Therapeutic genome editing: prospects and challenges. Nature Medicine 21(2):121–131. doi: 10.1038/nm.3793

Cyranoski D (2015) Embryo editing divides scientists. Nature 519(7543):272

Defensive drives (2015) Editorial. Nature 527:275–276. doi: 10.1038/527275b

Dickmann P, Sheeley H, Lightfoot, N (2015) Biosafety and biosecurity: a relative risk-based framework for safer, more secure, and sustainable laboratory capacity building. Frontiers in Public Health 3:241. doi: 10.3389/fpubh.2015.00241

European Commission (nd) Ethics. Horizon 2020: the EU framework programme for research and innovation. https://ec.europa.eu/programmes/horizon2020/en/h2020-section/ethics

Gantza VM, Jasinskieneb N, Tatarenkovab O, Fazekasb A, Maciasb VM, Biera E, James AA (2015) Highly efficient Cas9-mediated gene drive for population modification of the malaria vector mosquito Anopheles stephensi. Proceedings of the National Academy of Sciences of the United States of America 112(49):E6736–E6743. doi: 10.1073/pnas.1521077112

Hegg J (2016) Is intentional extinction ever the right thing? PLOS Ecology Community, 1 July. http://blogs.plos.org/ecology/2016/07/01/is-intentional-extinction-ever-the-right-thing/

Huang Y, Chen Y, Zeng B, Wang Y, James AA, Gurr GM, Yang G, Lin X, Huang Y, You M (2016) CRISPR/Cas9 mediated knockout of the abdominal-A homeotic gene in the global pest, diamondback moth (Plutella xylostella). Insect Biochemistry and Molecular Biology 75:98–106. doi: 10.1016/j.ibmb.2016.06.004

Ishii T (2015) Germ line genome editing in clinics: the approaches, objectives and global society. Briefings in Functional Genomics 16(1):45–56. doi: 10.1093/bfgp/elv053

Lanphier E, Urnov F (2015) Don't edit the human germ line. Nature 519:411

Ledford H (2015) CRISPR, the disruptor. Nature 522:20–24. doi: 10.1038/522020a

Leftwich PT, Bolton M, Chapman T (2016) Evolutionary biology and genetic techniques for insect control. Evolutionary Applications 9(1):212–230. doi: 10.1111/eva.12280

Mendell JR, Rodino-Klapac LR (2016) Duchenne muscular dystrophy: CRISPR/Cas9 treatment. Cell Research 26(5):513–514. doi: 10.1038/cr.2016.28

Next-generation genome editing (2015) Editorial. Nature Biotechnology 33(5):429. doi: 10.1038/nbt.3234

Oye, KA, Esvelt K, Appleton E, Catteruccia F, Church G, Kuiken T, Lightfoot SB, McNamara J, Smidler A, Collins JP (2014) Regulating gene drives. Science 345(6197):626–628

Reardon S (2014) Gene-editing method tackles HIV in first clinical test. Nature, 5 March. http://www.nature.com/news/gene-editing-method-tackles-hiv-in-first-clinical-test-1.14813. doi: 10.1038/nature.2014.14813

Reardon S (2016) First CRISPR clinical trial gets green light from US panel. Nature, 22 June. http://www.nature.com/news/first-crispr-clinical-trial-gets-green-light-from-us-panel-1.20137. doi: 10.1038/nature.2016.20137

Sovová T, Kerins G, Demnerová K, Ovesná J (2016) Genome editing with engineered nucleases in economically important animals and plants: state of the art in the research pipeline. Current Issues in Molecular Biology 21:41–62

Author Biography

Johannes Rath is head of the DNA laboratory in the Faculty of Life Sciences at the University of Vienna. Johannes works as an independent adviser to the Austrian government, the United Nations, and the European Commission.

Chapter 14
Seeking Retrospective Approval for a Study in Resource-Constrained Liberia

Jemee K. Tegli

Abstract The increase in the volume of health-related research in Africa has not everywhere been accompanied by improvements in research oversight systems related to biomedical and health services research, including the strengthening of institutional review boards (IRBs) and national regulatory oversight institutions. Critical to such oversight are not only competencies in ethics for the review of clinical trials, but also competencies in diverse research methods, statistical analyses and project implementation. In Liberia, there are recognized weaknesses in the existing infrastructure and capacity to regulate and monitor clinical research based on ethical institutional regulations and guidance for the protection of human research participants. During the height of the Ebola virus disease (EVD) surge in Liberia in 2014, there was a fragile national regulatory framework to oversee research. Some researchers took undue advantage of this gap to conduct unethical research.

Keywords Ebola Virus Disease · Liberia · Emergency research Retrospective approval · Ethics committee · Institutional review board

This case study is about an attempt to seek ethics approval *after* the study had already been conducted. Trying to obtain retrospective approval of research undermines the legitimacy of an Institutional Review Board (IRB) and the research review process itself. If the level of risk that a research study on EVD survivors presents is only determined after its completion, then the participants would have already been exposed to harm and their autonomy compromised. The researcher in this case used the cover of "emergency research" to avoid the review process – although emergency

J. K. Tegli (✉)
PREVAIL OFFICE, 1st Floor, East Wing, JFK Medical Center, 21 Street, Sinkor, Monrovia, Liberia
e-mail: tegli@ul-pireafrica.org

J. K. Tegli
UL-PIRE Africa Center, University of Liberia, Ground Floor, GD Bldg., Monrovia 100010, Liberia

D. Schroeder et al. (eds.), *Ethics Dumping*, SpringerBriefs in Research and Innovation Governance, https://doi.org/10.1007/978-3-319-64731-9_14

research regulations stipulate full disclosure of proposed research prior to implementation, and there are specified consenting processes (National Institute of Medicine (2002)).

Area of Risk of Exploitation

An investigator breached the Declaration of Helsinki and other global ethical guidelines by ignoring an ethical review oversight process that was fully functional in Liberia. The investigator was provided with information about the country's ethical review process at meetings of the Ebola Response Incident Management System, but chose to deploy the research team into targeted communities without ethical approval, collecting data through focus group discussions, key informant interviews and in-depth interviews. The data were analysed and submitted for presentation and/or publication at national and international meetings.

The entire data collection process, involving sensitive topics centred on Ebola-affected communities, was highly distressing. The consent process was conducted at the discretion of the researcher and the study team (Pollock, 2012). Providing so-called consent processes to participants in the frenzy of an epidemic can lead to misconceptions on their part. Risk mitigation measures were not fully assessed because guidance from an ethical institution was not explored or sought prior to the conduct of the study.

The fundamental pillar of respect for persons was also ignored. This should be the principle upon which each research participant makes an informed choice about whether or not to participate in the research, and thus accepts the potential risks and burdens of participation (Jasanoff, 1993). Thorough explanation to participants about the availability of the research ethics review board to address concerns regarding their rights or wellbeing, and providing contact information, are critical elements of the consent process. Whether consent is sought verbally or in writing, reference must be made to an ethics review board that reviewed and approved the research study. Ignoring one of the major pillars of ethical research – namely, ethics review – opens up a range of risks for exploitation (Martinson et al., 2005).

Background

Whatever the context, the need for the regulation of research is clear. In March 2014, the Ebola epidemic hit Liberia very hard. With Guinea and Sierra Leone also affected, the death toll in Liberia as of May 2016 stood at 4,810, with 10,678 infected. At the peak of the epidemic in October 2014, researchers and institutions were pouring into the country to conduct all forms of research, ranging from social science, anthropological and clinical studies.

As a result of the health emergency, the ethical research institutions were overwhelmed with questions from many investigators and institutions seeking information about the application process. The ethical boards were approached by several potential investigators in and outside of Liberia for information and guidance about the review process in the country. In spite of the availability of this platform, some researchers proceeded to conduct their research studies in November 2014 without prior review and approval by the relevant IRB. They circumvented the process under the pretext of the emergency and collected data from human participants in an unethical manner.

A particular case came to light in December 2014, when the IRB denied a researcher employed by an international UN public health agency retrospective ethical approval. Such an attempt to obtain retrospective approval of research undermines the legitimacy of the IRB and the research review process itself. If the level of risk that a research study on Ebola virus disease (EVD) survivors presents is determined only after its completion, then the participants would have already been exposed to harm and their autonomy compromised. This researcher used the cover of "emergency research" to avoid the review process – although emergency research regulations stipulate full disclosure of proposed research prior to implementation, and there are specified consenting processes.

Details of the Case

In early December 2014, the University of Liberia-Pacific Institute for Research and Evaluation IRB received an application from a researcher working for an international UN public health agency seeking approval for medical anthropological research on survivors of EVD in Monrovia. The focus of the study was on gathering information on the economic well-being of EVD survivors in Liberia. The objective was to assess the psychosocial situation and the impact of stigma and discrimination on the lives of survivors.

The research, which took place at the height of the EVD surge from November to December 2014, involved several EVD-affected communities in urban and peri-urban Monrovia. Most were from semi-literate and illiterate populations and had already been traumatized by the surge in EVD deaths. This study therefore had the potential to inflict distress on these participants: for example, it is traumatizing for EVD survivors to recount their experiences or circumstances, because these involve very recent catastrophic events in their lives.

The IRB convened and expedited the review process, given the importance of conducting the proposed research in a timely manner during the EVD outbreak. At the time, the IRB was not aware that the research had already been completed when the application was submitted for review. The researcher had left Liberia and was not present at the meeting, but was represented by a less experienced Liberian research assistant (RA).

During the IRB meeting, the RA was asked a question about the intended timetable for commencing the study. To the board's surprise, they were told that the data for the study had already been collected and analyzed. The role and responsibilities of this junior researcher in the implementation of the study were unclear and apparently minimal. The research team was only seeking approval from the IRB in order to disseminate the research results.

The IRB chair immediately called the meeting off and told the RA that the incident undermined best practices in the ethical conduct of health research. The RA was instructed to report the findings of the IRB to the principal investigator: namely that research had been conducted unethically, research review and approval regulations had been contravened, and the autonomy of the EVD survivors who were research participants had been breached.

Reasons for IRB Decision and Conclusion

A retrospective IRB approval of a research project – that is, after participants may already have been exposed to unnecessary harm or violation of their rights – would be unethical. Conducting a research study without IRB oversight violates both ethical principles and IRB procedures. In this context, the decision takes a stance on public policy that aims at increasing compliance with IRB requirements.

Without ethics approval, it is impossible to have the results of a study published in a reputable journal. One must assume that this is why the researcher tried to obtain retrospective approval. While the decision the IRB reached was emphatic, it was preceded by a short discussion on whether granting approval retrospectively, and thereby allowing publication of the research results, may contribute to the public good (Tansey et al., 2010). However, the IRB decided firmly in the negative, partly because the research had been undertaken on a highly vulnerable group of illiterate and uneducated Liberians who would have little knowledge of the consent process.

The Liberian IRB therefore made a clear decision to uphold the autonomy of both the IRB and the research participants by refusing to approve a study retrospectively.

References

Committee on Assessing Integrity in Research Environments (2002) Integrity in scientific research: creating an environment that promotes responsible conduct. The National Academies Press, Washington DC

Jasanoff S (1993) Innovation and integrity in biomedical research. Academic Medicine 68(9 Suppl):S91–S95

Martinson BC, Anderson MS, de Vries R (2005) Scientists behaving badly. Nature 435 (7043):737–738

Pollock K (2012) Procedure versus process: ethical paradigms and the conduct of qualitative research. BMC Medical Ethics 13(1):25. doi: 10.1186/1472-6939-13-25

Tansey CM, Herridge MS, Heslegrave RJ, Lavery JV (2010) A framework for research ethics review during public emergencies. Canadian Medical Association Journal 182(14):1533–1537. doi: 10.1503/cmaj.090976

Author Biography

Jemee K. Tegli is Director of Operations for the Liberia-United States Clinical Research Program, Partnership for Research on Ebola Virus in Liberia. He previously worked for the University of Liberia-Pacific Institute for Research and Evaluation as Center Director. He serves as Coordinator of the UL-PIRE Institutional Review Board (IRB) and Member of the National Research Ethics Board (NREB) of the Ministry of Health. He is an international IRB Member of the Western Intuitional Review Board (WIRB)-Copernicus Group in Puyallup, Washington.

Chapter 15
Legal and Ethical Issues of Justice: Global and Local Perspectives on Compensation for Serious Adverse Events in Clinical Trials

Yali Cong

Abstract A 78-year-old Chinese woman joined a clinical trial sponsored by a pharmaceutical company. Unfortunately a serious adverse event (SAE) occurred. The sponsor paid for the cost of the medical care arising from the SAE, but refused the family's request for compensation. The family then sued the company and the hospital in Beijing. Although the SAE was related to a complication of lower extremity angiography and not the drug itself, it was a direct consequence of participating in the trial. According to Good Clinical Practice, a set of regulations promulgated under Chinese law, "the sponsor should provide insurance to those human subjects who participate in clinical trials, cover the cost of treatment and the corresponding economic compensation for the occurrence of the harm or death associated with the trial" (SFDA in Good clinical practice. State Food and Drug Administration,2003: art. 43). The court ordered the trial sponsor to provide a translation of the company's insurance policy, so that the court could understand the amount of compensation available to the patient under the policy, but the sponsor never surrendered either the documentation or a translation. Consensus was never reached about the amount of compensation due to the patient through negotiation with the hospital, the company and the family. The litigation ended after nine hearings and five long years. This chapter provides an ethical analysis of the case relative to at least three areas of risk of exploitation when a major, international pharmaceutical company sponsors clinical research in a country with an immature legal system and where research participants have limited resources.

Keywords Clinical trial · Serious Adverse Event (SAE)
China · Global research · Justice · Insurance

Y. Cong (✉)
Medical Ethics and Health Law Program, Peking University Health
Science Center, 38 Xueyuan Rd, Beijing, China
e-mail: ethics@bjmu.edu.cn

© The Author(s) 2018
D. Schroeder et al. (eds.), *Ethics Dumping*, SpringerBriefs in Research
and Innovation Governance, https://doi.org/10.1007/978-3-319-64731-9_15

Areas of Risk of Exploitation

There are at least three ways in which this case illustrates the risk of exploitation.

The principle of justice requires that the benefits and burdens of research be distributed fairly. This means that participants who are injured during the research should be compensated fairly for their injuries. The present case demonstrates the main risk of exploitation during the process of an individual research participant's litigation. Although individuals may be compensated, litigation is costly and time-consuming. Studies have found that approximately 50% of the sums recovered from tort lawsuits in high-income countries (HICs) do not reach the injured parties but instead go to attorney fee payments and other costs. Legal barriers such as the assumption of risk, contributory negligence and government immunity may discourage litigation by injured research participants or preclude recovery in whole or part (Resnik et al. 2014).

Second, this case illustrates the risk of exploitation due to the considerable variation in regulations across various countries, which results in inconsistent compensation for the victim of a serious adverse event (SAE). Regarding the payout amount for compensation, trial sponsors might approach the amount differently for human research participants who suffer the same SAEs in different countries. This suggests that the values of justice may not be fulfilled, as there should be no double standards in the compensation for SAEs. While there is no data publicly available about variations in payment for SAEs, this case raises the suspicion that equal and just compensation in global studies is not being achieved, or at least not in all cases. Exploitation occurs when different patients suffer the same harm or injury, but do not receive equal compensation (or at least compensation adjusted to amounts based on average incomes in the countries concerned).

The third risk of exploitation derives from the inequality in access to resources for litigation between individual research participants and pharma sponsors. In this case the company exploited its position of litigatory strength. It did not cooperate with the local court, in that, for example, it did not supply either the original of the insurance contract or a translation into Chinese. In addition to being a failure to comply with the court's request, this delayed the legal process.

Background

Xarelto® (rivaroxaban) or BAY 59-7939 is an oral tablet (factor Xa inhibitor), taken once a day, intended for prophylaxis of deep vein thrombosis (DVT) and the prevention of atrial fibrillation, cardiac thromboembolism and cerebral infarction. The company's application "On the BAY 59-7939 international multi-centre phase III clinical trial" was submitted to China's State Food and Drug Administration

(SFDA) in October 2005 and approved in February 2006. The institutional review board (IRB) of a hospital approved the trial based on the application. Being a global clinical trial, the hospital was invited as the leading centre. The trial sponsor signed a contract with the hospital.

The Case

A 78-year-old woman came to hospital for knee replacement surgery. During her index hospitalization in 2006, she was invited to join this clinical trial. Her daughter was with her at the time of recruitment, and they both agreed to her participation. The knee replacement surgery was conducted on 24 October 2006. In accordance with the protocol, she took the daily tablet, intended for prophylaxis of DVT and the prevention of atrial fibrillation, cardiac thromboembolism and cerebral infarction. She was enrolled from 23 October to 6 November 2006. The research protocol required the patient to undertake double lower-limb vein angiography in order to test for thrombus formation. An SAE occurred after venous angiography. The patient suffered chest tightness, shortness of breath, palpitations, cough, sweating, a very weak pulse, blood pressure dropping to 60/40 mm HG and shock. The patient regained consciousness three hours after resuscitation. The hospital's principal investigator judged this complication to be an SAE and completed the SAE report form on 15 February 2007. The SAE was also reported to the China State Food and Drug Administration on the same day.

The total expenses of the medical treatment caused by the SAE were CNY 3296.17 (approximately USD 420 in 2006), all of which the trial sponsor paid. Considering the patient's suffering and the adverse effect on her recovery of knee function, she and her family desired compensation for the limitations the SAE had imposed on her life. The patient and her family knew that the sponsor had compensation insurance for the study. The investigator reminded the patient of this, and they found relevant information in the informed consent form.

In the section entitled "Patient Notice", the consent form read, "if a subject involved in this trial is injured during the study, the insurance company will pay correspondingly". Based on the consent form, and the study investigator's explanation, the patient knew that the trial sponsor had purchased global insurance for this multi-centre clinical trial. When the patient and her family requested compensation from the hospital and the pharmaceutical company, the company refused. Despite extensive discussions, the three parties could not reach consensus. After failure to agree on a compensation amount, the sponsor and the hospital were summoned to the Beijing Chaoyang Court by the plaintiff in 2008.

Procedure for Compensation Claim in China

Usually in China, if a plaintiff is injured and claims compensation, the court will require the plaintiff to consult with a third party to evaluate the nature and degree of the injury. Based on this evaluation, the court can then make a judgement about the seriousness of the injury, and determine an amount of compensation. In this case, the children did not want to expose their elderly mother to the pressure of visiting the evaluation centre and having to wait a long time for a result that she might not be satisfied with anyway, so they decided to spare their elderly mother this ordeal. The family did not file a suit as a lawsuit based on infringement of rights or as a suit of tort, but filed as a "dispute of contract".

The Source of Disagreement

The sponsor argued that there was an agreement between themselves and the hospital to carry out the trial of a new drug, and hence a contractual relationship between the company and the hospital, but there was no contractual relationship between the company and the patient plaintiff. In contrast, the plaintiff argued that the hospital had clearly informed the patient (research participant) that the hospital was only a representative of the trial sponsor, and further that the research participant had been informed that the company had entrusted/endorsed the hospital to sign the contract with trial participants. On these grounds, the plaintiff declared that a contractual relationship existed between the plaintiff and the company. The Chaoyang Court ultimately accepted the plaintiff's claim as a dispute of contract.

Having accepted the suit, the court requested the parties to provide the relevant documents. It repeatedly requested the company to provide a copy of the insurance contract, and explained this requirement to the company, but the company resisted and did not submit the insurance contract. The court also asked the hospital for the insurance contract, but the hospital responded that it had been unaware that it should request that documentation. Similarly, the hospital ethics committee had not required confirmation of an insurance contract at the time the protocol was approved. The hospital argued that it had signed a clinical trial contract with the company, which had declared that it had purchased special insurance to cover economic loss by the subjects participating in the study, including any harm caused by the drugs. The third page of the participant information sheet for the study stated: "Adverse drug reactions related to angiography include angiography reaction, such as skin reaction; some will imply allergic reaction, such as anaphylactic shock". Thus an adverse event from the double lower-limb vein angiography was included.

The plaintiff then requested the hospital ethics committee to seek help from the SFDA, but the committee were informed that the SFDA did not have this document either. In short, no one but the company had access to a clear description of the amount of compensation during the earlier stages of the case.

Later in the court process, the pharma company provided Chaoyang Court with certification of insurance purchased from a German provider, certifying that the company purchased insurance effective from January 1, 2002, which covered the study overseas and participants from all countries. Each person's maximum insurance was approximately 500,000 Euro. (Chao Min Chu Zi 2009).

The court asked the company to provide a Chinese version of the insurance contract, but the company refused. After several requests, the court, the plaintiff and her family were informed that it would take a long time to prepare such a translation and that it would be too expensive (estimated cost CNY 20,000, approximately USD 3,000 at that time).

This meant that the available documentation – namely, the consent form and insurance contract – included no clear description of the exact amount of compensation, nor how to compensate for different situations, types of injury, different countries, etc. When the plaintiff claimed EUR 150,000 compensation, the company argued that there was no reasonable basis for such a claim.

Though no specific criteria were provided about the amount of compensation, the civil judgement included a clause referring to the insurance company's view that where the company was responsible for the compensation of subjects, the insurer should provide the compensation based on the requirements of the local laws where the injury occurred (Chao Min Chu Zi 2009). The pharmaceutical company requested a non-public hearing for the appeal, which made information unavailable. The plaintiff explained that she was persisting with her appeal as she suspected that there was an unfair clause in the insurance contract and that there was an unequal description of compensation for HICs and LMICs.

After five years, the lower court's judgement was issued in February 2013. The Beijing Chaoyao Court determined that according to the Chinese Good Clinical Practice regulations, the company should compensate the plaintiff with EUR 50,000. However, the plaintiff did not accept this, and appealed to the Beijing Second Middle Level Court. That court rejected the appeal.

In summary, between 2009 and 2011, nine hearings were held. The final conclusion came out in February 2013. The entire process of litigation and appeal lasted for five years. Compensation of EUR 50,000 euros was paid directly by the company, not by the insurance company. This suggested that the process of SAE compensation was dealt with internally within the company, rather than through a formal procedure that involved the insurance company.

Due to the SAE and consequent extended hospitalization, the patient was placed on strict bed rest, even though rehabilitation from the original knee replacement surgery would have required her to move. Her dream had been to travel abroad after the surgery, but participating in the trial delayed her rehabilitation from the surgery.

Update

It was reported from Berlin on 4 May 2015, that the company's once-daily oral anticoagulant BAY 59-7939 (rivaroxaban) had been approved by China's State Food and Drug Administration for the prevention of stroke and systemic embolism in adult patients with non-valvular atrial fibrillation with one or more risk factors (Bayer 2015). Additionally, the administration has approved BAY 59-7939 for the treatment of DVT and the reduction of the risk of recurrent DVT and pulmonary embolism following acute DVT in adults. Since 2009, BAY 59-7939 has been available in China for the prevention of venous thromboembolism in adult patients undergoing elective hip or knee replacement surgery.

Lessons Learned & Recommendations

- Though the capacity for human research participant protection and ethics review have been improved in China in recent years, this case shows that some matters may have been neglected, especially access to the insurance contract for compensation. In this case, all three stakeholders should strengthen their sense of responsibility and learn this lesson: the hospital's ethics review committee did not fulfil its responsibility to request the company to provide the insurance contract. The SFDA needs to develop a working system which ensures that the pharmaceutical company sponsoring a trial prepares and submits to the ethics review committee relevant documents such as the insurance policy as a requirement.
- While both the local Chinese Good Clinical Practice regulations and the Guideline for Good Clinical Practice of the International Council for Harmonisation of Technical Requirements for Pharmaceuticals for Human Use (ICH 1996) have provisions about compensation, it is hard for an individual research participant to negotiate and reach consensus with individual companies. For example, article 43 of the Chinese GCP (SFDA 2003) addresses compensation, but clearly places the responsibility on the shoulders of the sponsor.
- The prolonged processes involved in the interpretation and application of the law also contribute to potential harm and exploitation of trial participants and their families. In this case, the lawsuit started in 2008 when the plaintiff was 79 years old and ended when she was 85. Her dreams of travel after the knee replacement surgery were shattered.
- Bringing a legal case always involves costs for the plaintiff, which have to be advanced at least until the court reaches its finding or insurance is paid. It is often impossible for vulnerable populations in research to provide fees to lawyers and courts. (This case was an exception.)

- Within China, as this case illustrates, an academic dialogue is needed on the nature of the relationship between individual human research participants and a trial sponsor. During medical treatment, patients and doctors form a fiduciary relationship, as well as a contractual relationship. There is academic discussion of the doctor-patient relationship. However, there is not yet an academic discussion about the nature of the relationship between research participant and trial sponsor.
- This case calls into question whether compensation for injury should be a set amount, an amount based on an individual's economic situation, or an amount based on a country's economic situation. Regarding the amount of compensation to an individual research participant with an SAE during a global clinical trial, ethicists need to address the ethical challenge of a double standard.
- One final lesson relates to the exploitation of a less mature legal system. China, like many other middle-income countries, lacks lawyers and legal teams who are able to provide support in litigation with a pharmaceutical giant.

Acknowledgements   To Liao Zhijie, the son of the plaintiff in this case: he emailed me the case materials and expressed willingness to have this used as a training case.

To lawyer Gao Mei, who was once the attorney of the plaintiff: I discussed the preparation of this case study with her, and she exchanged some important points with me.

To Professor Michael Fetters,University of Michigan, who was Fulbright Distinguished Chair in the Social Sciences, Peking University Health Science Center, China, Sept 2016–Jan 2017: he provided many valuable points and pointed out some language inaccuracies.

References

Bayer (2015) Bayer's Xarelto® approved in China for stroke prevention in patients with non-valvular atrial fibrillation and for the treatment of deep vein thrombosis. http://www.investor.bayer.de/index.php?id=145&L=1&tx_news_pi1[news]=1808

Chao Min Chu Zi (2009) Beijing Chaoyang District people's court, civil judgments, No. 02608. Blog. SINA. http://blog.sina.com.cn/s/blog_c2b037dd0101f29q.html

ICH (1996) Guideline for good clinical practice. International Council for Harmonisation of Technical Requirements for Pharmaceuticals for Human Use, Geneva. http://www.ich.org/fileadmin/Public_Web_Site/ICH_Products/Guidelines/Efficacy/E6/E6_R1_Guideline.pdf

Resnik DB, Parasidis E, Carroll K, Evans JM, Pike ER, Kissling GE (2014) Research-related injury compensation policies of U.S. research institutions.IRB: Ethics & Human Research 36 (1):12–19. https://www.ncbi.nlm.nih.gov/pmc/articles/PMC3991013/

SFDA (2003) Good clinical practice. State Food and Drug Administration, People's Republic of China

Author Biography

Yali Cong is professor of medical ethics at Peking University's Health Science Centre. In 2010, she was involved in establishing the university's first human subject protection programme. In 2008, she set up a joint centre for medical professionalism with the University of Columbia, and serves as deputy director. She is currently vice chair of the China Medical Ethics Association.

Other Resources

These open access casebooks may be useful to the reader:

Barrett DH, Ortmann LW, Dawson A, Saenz C, Reis A, Bolan G (2016) Public health ethics: cases spanning the globe. Springer International Publishing. http://www.springer.com/us/book/9783319238463

Cash R, Wikler D, Saxena A, Capron A (2009) Casebook on ethical issues in international research. World Health Organization, Geneva. http://www.who.int/rpc/publications/ethics_casebook/en

UNESCO (2011) Casebook on benefit and harm. Bioethics core curriculum. Casebook series no. 2. UNESCO, Paris. http://unesdoc.unesco.org/images/0019/001923/192370e.pdf

UNESCO (2011) Casebook on human dignity and human rights. Bioethics core curriculum. Casebook series no. 1. UNESCO, Paris. http://unesdoc.unesco.org/images/0019/001923/192371e.pdf

The following high-calibre Ph.D. thesis contains a range of excellent case studies:

Ravinetto R (2016) Methodological and ethical challenges in non-commercial North-South collaborative clinical trials. Leuven University Press, Leuven

Table 1 shows how the case studies selected for this book align with the ethical issues identified by the Horizon 2020 programme of the European Commission.

D. Schroeder et al. (eds.), *Ethics Dumping*, SpringerBriefs in Research and Innovation Governance, https://doi.org/10.1007/978-3-319-64731-9

Table 1 Ethics dumping cases organized according to Horizon 2020 ethical issues checklists

Ethical issue	Subcategory	Chapter
Human embryos and foetuses		None identified in our research
Humans	Volunteers for social science research	2
	Persons unable to give consent, including minors	6
	Vulnerable groups	2, 3, 4, 5, 6, 8, 9
	Patients	15
	Healthy volunteers in medical studies	4, 5, 6, 7, 8
Human cells and tissues		4
Personal data		12
Animals		10
Third countries: *Not applicable, as all cases in this report involve third countries*		
Environment, health and safety	Environment	11, 13
	Health and safety	11, 13
Dual use		13
Misuse		12, 13

http://ec.europa.eu/research/participants/data/ref/h2020/grants_manual/hi/ethics/h2020_hi_ethics-self-assess_en.pdf

Index

A

Adverse Events (AE), 7, 62, 124
Africa, 4, 10, 16, 17, 20, 24, 26, 49, 50, 52, 85, 86, 95, 99, 115
Agriculture, 93, 108, 110
Anhui province, 71–74
Anthropological research, 10, 117
Autonomy, 28, 63, 64, 115, 117, 118

B

Banana, 6, 91, 93–96
Bantu, 4, 15, 16, 18, 20
Benefit, 2, 5, 10, 12, 13, 16, 21, 27, 45, 50, 52, 55, 57, 61–65, 67, 69, 71, 72, 75–78, 82, 83, 85, 100–102, 107, 109, 122
Benefit sharing, 4, 5, 28, 51
Bill and Melinda Gates Foundation, 35, 39, 93
Biomedical research, 44, 51, 82–85, 87, 88, 100
Biotechnology, 82
Blood samples, 5, 57, 71, 73, 74
Botswana, 16, 20
BUAV, 85
Bushmen, 17, 18, 20

C

Canada, 5, 23, 28, 49, 58, 84
Candidate vaccine trial, 50
Capacity building, 28
Cartagena protocol, 111
Cell-Life, 103
Cervical cancer, 4, 34–43
Children, 10, 24, 49, 50, 52, 62, 95, 124
China, 5, 72–77, 84, 86, 122–124, 126, 127
Civil society, 52
Clinical trial, 4, 5, 7, 33, 50–53, 58, 61–65, 67–69, 109, 115, 123, 124, 127
Cluster Randomised Controlled Trial (CRCT), 36, 40

Community, 4, 9–13, 16, 19, 21, 26–29, 33, 36, 37, 39, 44, 50, 52, 53, 71, 73, 77, 78, 84, 103, 108
Community-based studies, 103
Community consent, 13, 19
Community consultation, 21
Compensation, 7, 45, 56, 64, 65, 78, 122–127
Compliance, 3, 57, 87, 118
Confidentiality, 13, 63–65, 69, 101
Conflicts of interest, 53
Consent, 5, 9, 10, 13, 16–19, 26, 28, 42, 44, 51–54, 56, 63, 67, 77, 92, 94, 100, 101, 103, 104, 116, 118, 123, 125
Contraception, 55, 62
Contract Research Organisation (CRO), 68
Control arm, 33, 35, 37–40, 42–45
Control group, 36, 40, 41, 44
Convention on the elimination of all forms of discrimination against women, 64
Council for International Organizations of Medical Sciences (CIOMS), 37, 87
CRISPR/Cas9, 6, 107
Culture, 9–11, 19, 86, 96
Cytology screening, 33, 36, 38

D

Data, 4, 5, 9, 11–13, 16, 18, 39, 44, 45, 63, 67, 68, 75, 87, 100–104, 116–118, 122
Data protection, 63, 69, 104
Declaration of Helsinki, 37, 44, 51, 53, 55, 63, 64, 68, 116
Diagnostic laboratory, 49, 51
Diarrhoea, 10, 11, 53
Discrimination, 16, 64, 117
DNA, 16, 19, 35, 40, 74, 75
Double standards, 2, 7, 34, 43, 44, 51, 122, 127

E

East Africa, 24, 93, 95

© The Editor(s) (if applicable) and The Author(s) 2018
D. Schroeder et al. (eds.), *Ethics Dumping*, SpringerBriefs in Research and Innovation Governance, https://doi.org/10.1007/978-3-319-64731-9

Ebola, 5, 50, 53–56, 58, 116, 117
Emergency research, 6, 7, 115, 117
Epidemic, 5, 25, 50, 54, 116
Epidemiology, 26
Ethical guidelines, 37, 44, 116
Ethical principles, 38, 42, 63, 118
Ethical review, 3, 11, 13, 61, 65, 76, 77, 116
Ethical standards, 7, 9, 11, 33, 34, 43, 76, 78,
    81, 104
Ethics approval, 3, 6, 65, 115, 118
Ethics committee, 2, 15, 19, 20, 44, 45, 49,
    51–53, 56, 57, 75, 87, 96, 124
Ethics dumping, 1, 2, 7, 33–35, 42, 43, 45, 51
Ethics Review Committee (ERC), 10, 57, 126
European Commission, 84, 108
European Commission's Scientific Committee
    on Health, Environmental and Emerging
    Risks (SCHEER), 84
European Union, 6, 81
European Union Directive 2010/63/EU, 81, 83,
    84, 87
Exclusion criteria, 36, 83, 125
Experimental, 39, 43, 58, 78, 82, 87, 108
Exploitation, 1–5, 9, 10, 12, 16, 20, 21, 23, 28,
    34, 50, 51, 61, 64, 67, 71, 72, 77, 78, 81,
    87, 88, 91, 92, 100, 102, 104, 108, 111,
    121, 122, 126
Exploitative, 1, 15–17

F
Farmers, 73–76, 92
Female Genital Mutilation (FGM), 4, 9, 11, 12
Financial incentives, 5, 57, 58, 67, 96
Food security, 6, 91, 95
Food sovereignty, 91, 96

G
Gender, 5, 61, 64, 65
Genetically Modified (GM), 6, 91, 93–96
Genetically Modified Organisms (GMO), 94,
    110
Genetic research, 5, 75
Girl, 11, 12, 25
Golden rice, 95
Good Clinical Practice (GCP), 61, 63, 121,
    125, 126
Government, 5, 12, 26, 33, 35, 36, 39, 42–44,
    49, 51, 71, 72, 74, 75, 77, 102, 122
Guideline for Good Clinical Practice of the
    International Council for Harmonisation
of Technical Requirements for
    Pharmaceuticals for Human Use (ICH
    1996), 126

H
Harm, 6, 12, 20, 35, 42–45, 68, 100, 108, 115,
    117, 118, 122, 124, 126
Health emergency, 117
Health workers, 24, 39, 40
Healthy volunteers, 5, 50, 58, 62, 67–69, 96
Hepatitis B, 5, 62, 63
High Income Countries (HIC), 2, 6, 34, 36, 42,
    44, 45, 67, 73, 76–78
HIV, 23, 26, 27, 29, 103, 109
HIV/AIDS, 4, 26, 103
Horizon 2020, 84
Humanitarian, 4, 10, 12
Human Papillomavirus (HPV), 35–37, 40–42
Human participants, 44, 59, 91, 117
Human rights, 11, 29, 35, 42, 57, 59, 64, 78
Hunger, 78, 91, 94
Hunter gatherers, 4, 15–20

I
Illegal, 4, 9, 11, 23, 27, 44, 77
Illiterate populations, 117
Immunobiology, 26
Incentives, 57, 58, 67, 96, 102
Inclusion criteria, 36, 83, 125
India, 4, 35–44, 68
Indigenous people, 17, 18
Inequality, 73, 94, 122
Information sheets, 53, 124
Informed consent, 5, 12, 13, 15–19, 27, 28, 35,
    39, 42, 44, 52–56, 59, 61, 63, 64, 67, 69,
    75, 92, 94, 101, 123
Informed consent process, 15, 17, 53
Institutional Review Boards (IRBs), 19, 39,
    115, 117, 118, 123
Insurance, 44, 63, 64, 121–126
International Council for Laboratory Animal
    Science and the Council for
    International Organizations of Medical
    Sciences (ICLAS), 87

K
Kenya, 23–25, 27–29, 85, 86
Khoisan, 15, 17, 18

L

Legal framework, 6, 57, 86, 107, 108
Lesbian, Gay, Bisexual and Transgender persons (LGBT), 24
Liberia, 6, 54, 115–117
Litigation, 121, 122, 125, 127
Local community, 1, 9, 10
Local Ethics Committee (LEC), 61, 65
Low and Middle Income Countries (LMIC), 2, 34–37, 41–43, 45, 50, 54, 57, 59, 68, 76–78, 91, 96, 100

M

Majengo, 23–29
Majengo Observational Cohort Study (MOCS), 26, 29
Medical research, 9, 16, 44, 51, 59, 63, 64, 73, 85, 100
mHealth, 6, 99–104
Multi-country, 49
Multi-site, 49
Mumbai, 3, 33, 37, 38, 40, 44
Mxit, 103

N

Nairobi, 4, 23–26, 29, 30, 85
Namibia, 4, 15, 16, 20
National Ethics Committee (NEC), 49, 50, 52
Nature journal, 19
Non-compliance, 5, 61
Non-Governmental Organisations (NGO), 3, 4, 9–13, 93
Non-human primates, 6, 81–86, 88
Nutrition, 91
Nutritionism, 92–94

O

Osmanabad, 33, 37–41, 43, 44

P

Pap smear, 4, 33, 35, 38, 40, 43
Patents, 76, 78
Patient, 35, 54, 63, 99–103, 121–127
Peer educator, 27, 28, 30
Pharmaceutical companies, 5, 7, 49, 50, 56, 69, 71, 75, 76, 121, 123, 125, 126
Phase I/II clinical trial (testing for safety and immunogenicity), 5, 49, 50
Phase III clinical trials probably separately, 122
Philippines, 95
Placebo, 34, 35, 37, 44, 50, 54
Pre-Exposure Prophylaxis (PrEP), 28, 29
Pregnancy, 54, 55, 62, 63

Privacy, 64, 65, 78, 100–104
Proxy consent, 54
Publication, 15–18, 21, 56, 93, 108, 116, 118
Public awareness, 77, 78

R

Randomized double blinded trial, 49, 50, 61
Recruitment, 49, 50, 52–54, 59, 123
Regulatory approval, 4, 42, 44, 51, 57, 61, 76–78, 84, 115
Reproductive rights, 63–65
Research and Development (R&D), 73, 74, 77, 78, 91
Research ethics, 5, 18, 19, 21, 27, 42, 43, 51, 52, 71, 74, 75, 78, 116
Research participants, 2, 5, 9, 10, 27, 28, 33, 37, 43, 49–52, 55–59, 71, 73, 76, 77, 115, 116, 118, 121, 122, 124, 126, 127
Research review process, 115
Respect, 15, 19, 20, 35, 40, 57, 59, 75, 78, 84, 116
Risk mitigation measures, 116
Risks, 2, 3, 5, 9–13, 16, 26, 33, 34, 42, 50, 51, 53, 55, 64, 65, 67–69, 72, 84, 88, 91–96, 100, 101, 104, 107–111, 115–117, 121, 122, 126
Rural, 10, 16, 18, 25, 33, 35, 38, 40, 72, 95
Russia, 5, 61–63
Russian Federation, 61, 63

S

Safety, 5, 6, 23, 44, 49, 50, 52, 56, 61, 62, 68, 75, 82, 83, 94, 100, 104, 107–111
Sample, 5, 16, 26, 37, 39, 40, 49, 56, 57, 62, 73–77
San, 4, 15–21
San Code of Research Ethics, 21
Science & Technology (S&T), 71–73, 76–78
Semi-literate populations, 117
Serious Adverse Event (SAE), 7, 62, 63, 121–123, 125, 127
Sex workers, 4, 23–30
Sex Workers Outreach Programme (SWOP), 23, 27, 28, 30
Side effects, 50, 55
Sierra Leone, 54, 116
Slum, 4, 23, 24, 30, 37, 38
Social science research, 4, 116
Socio-anthropological research, 10
South Africa, 16, 20, 21, 28, 86, 103
South African San Council, 17
South African San Institute (SASI), 17
Southern Africa, 4, 15–17

Sponsor, 3, 5, 7, 42–45, 51, 61, 62, 122–124, 126, 127
Standard of care, 4, 33, 34, 37, 38, 40, 42, 43, 45
Standard regulatory procedures, 51
Stigmatization, 4
Stigmatized, 4, 12, 23, 24
Sub-Saharan Africa, 49, 50

T
Tradition, 84
Traditional healers, 10, 11
Training, 36, 52
Transgenic, 6, 91–93
Transparency, 45, 52, 57, 59, 93
Treatment, 3, 11, 25–27, 35, 37–42, 44, 45, 50, 53, 56, 64, 73, 86, 103, 107–109, 121, 123, 126, 127
Trial, 4–6, 34–45, 50–53, 55, 58, 61, 62, 65, 68, 92–94, 96, 122–126
Trust, 3, 5, 12, 21, 30, 49, 58, 59, 108, 109

U
Uganda, 6, 91, 93, 95, 96
UK, 85
UNICEF, 11
United Nations, 64
Universal Declaration of Human Rights, 64

Universal Declaration on Bioethics and Human Rights (UNESCO), 64
Urban, 25, 33, 35, 50, 72, 117
US government's Office of Human Research Protections (OHRP), 39, 44
US National Institutes of Health, 35, 39, 74, 84

V
Vaccine, 5, 26, 42, 50, 53–56, 58, 59, 61–63, 82
Vaccine trial, 5, 50, 56, 58
Vulnerability, 10, 52, 64, 92
Vulnerable, 2, 4, 10, 12, 13, 16, 23, 24, 30, 34, 35, 38, 42, 64, 67, 69, 104, 126
Vulnerable groups, 63, 67, 118

W
West Africa, 5, 49, 50
WHO, 26, 35, 36, 53, 95, 99
Women, 4, 5, 11, 24–27, 29, 33, 35, 37–43, 45, 52, 61–65, 92–94
Working Group of Indigenous Minorities in Southern Africa (WIMSA), 17, 19
World Bank, 92
World Trade Organization (WTO), 34

Z
Zika, 110